Biotechnology

SUNY series in Philosophy and Biology

David Edward Shaner, editor

BIOTECHNOLOGY

Our Future as Human Beings and Citizens

Edited by

SEAN D. SUTTON

Published by
STATE UNIVERSITY OF NEW YORK PRESS, ALBANY

Printed in the United States of America

For information, contact State University of New York Press, Albany, NY
www.sunypress.edu

Production by Robert Puchalik
Marketing by Fran Keneston

Library of Congress Cataloging-in-Publication Data

Biotechnology : our future as human beings and citizens / edited by Sean D. Sutton.
p. cm. — (SUNY series in philosophy and biology)
Includes bibliographical references and index.
ISBN 978-1-4384-2685-3 (hardcover : alk. paper)
ISBN 978-1-4384-2686-0 (pbk. : alk. paper)
1. Biotechnology—Social aspects. 2. Bioethics. 3. Biotechnology—Religious aspects.
I. Sutton, Sean D., 1965–
TP248.23.B5626 2009
303.48'3—dc22

2008043249

10 9 8 7 6 5 4 3 2 1

For

Sybil, Jonathan, Madeleine Sophie,

Michael and Charles

Contents

Acknowledgments

Biotechnology: Our Future as Human Beings and Citizens grew out of a lecture series sponsored by the Department of Political Science at Rochester Institute of Technology during the 2004–2005 academic year. The Biotechnology series was the main academic event of RIT's 175th anniversary celebration. The chapters in this book, with the exception of those by Ronald M. Green, Richard Sherlock, and Hava Tirosh-Samuelson, were originally delivered as part of that series.

I wish to thank Albert J. Simone, emeritus president of RIT, and Andrew M. Moore, former dean of the College of Liberal Arts, for the financial support and enthusiasm that made both the series and this volume possible. Earhart Foundation and the New York Council for the Humanities, the state affiliate of the National Endowment for the Humanities, also provided essential funding.

Finally, I wish to thank John A. Murley, professor of political science at RIT, and Larry Arnhart, professor of political science at Northern Illinois University, who generously insisted that I put this book together and see it through to publication.

Sean D. Sutton

INTRODUCTION

Biotechnology, Human Being, and Citizen

SEAN D. SUTTON

I

This book is a collection of essays on the recent advances in biomedical science and technology. The authors reflect upon the challenges posed by biotechnology: of securing the good that is motivated by the desire to cure disease and relieve human suffering and by the desire to respect life, and human dignity. Each author is well versed in the debate and represents a broad spectrum of opinion ranging from the cautionary to the celebratory, from the religious to the Darwinian. Such variety gives rise to a fundamental controversy among its contributors that reaches the heart of the issue of whether biotechnology is a threat to humanity or whether it is a boon. Readers will also find that the contributors' evident familiarity with each other's work binds together the essays into a coherent dialogue that mirrors the wider conversation engaging the citizens and representatives of this country.

Leon R. Kass, Richard Sherlock, and Hava Tirosh-Samuelson present the case for caution and regulation. They warn that the nontherapeutic use of biotechnology can remake human nature and debase humanity. Ronald Bailey, Ronald M. Green, and Lee M. Silver present the optimistic case that biotechnology should be embraced and encouraged. They present the case for a moral imperative to use biotechnology to eradicate disease and disability, and improve upon our natural endowment. Larry Arnhart's essay cuts across both sides of this debate. He questions the alleged claim that biotechnology can alter human nature, arguing that its uses will be limited both in its moral ends and in its technical means by man's natural moral sense and the complexities of human nature itself.

Taken together these essays constitute a profound inquiry into the promises and concerns inspired by biotechnology and biomedicine. This book leads the reader to the central issues of the debate. As a self-governing people, we will ultimately answer the questions surrounding biotechnology through our representatives and in our courts. We offer *Biotechnology: Our Future as Human Beings and Citizens* as a suitable primer for such a people.

II

The promises of stem cell research, of gene therapy, of psychotropic pharmacology have seized our imagination and have fueled our hopes for healthier bodies, happier souls, and longer lives. Yet, it is hard to deny that with every biotechnical breakthrough, with every potential cure for disease or affliction, there is an accompanying sense of unease. The promised benefits of biotechnology simultaneously provoke both the hope that our science can overcome human frailty by remaking ourselves and our children and the misgiving that this power is a threat to our very humanity.

Leon R. Kass points out that it is obvious to all that we stand at the threshold of the golden age of biomedicine and biotechnology. He acknowledges that we should be grateful that we do, as there is much to be thankful for, yet he notes that there is much about which to be concerned. His concerns are based in the recognition that the powers made possible by biomedical science can also be used for nontherapeutic purposes, serving ends that range from the frivolous and disquieting to the offensive and the pernicious. For example, Kass argues that drugs that improve memory, alertness, amiability, and sports performance are artificial means that endanger virtue, the need for strength of character, and the discipline so necessary for self-improvement. These artificial means may improve us, or at least we feel something that we can call improvement, but we do not play any role in the improvement. What is at stake, therefore, is the texture and richness of human life.

Ron Bailey recognizes that the opposition to biotechnology has brought together a strange coalition of parties, a fusion of left-wing environmentalists, religious fundamentalists, and neoconservatives. What draws them together, Bailey notes, is the fear that the widespread use of biotechnology will diminish our humanity and bring about an era of "posthumanity," where we will be both greater and lesser than ourselves. Bailey contends that these fears are overblown. The technology is not there yet to transform human nature, and there is nothing to fear from healthier bodies and longer lives.

Ronald M. Green adds that the growing opposition to biomedical research has undermined the dream to progressively free human life from the ancient scourges of disease and disability. Green argues this growing opposition has heralded a new era marked by a loss of nerve. Scientists have been hamstrung in their research efforts by legal prohibitions and government regulation that have produced a hostile atmosphere that will drive away young researchers and investors. His fear is that we will miss the benefits of the biotechnological revolution and unnecessarily suffer from diseases and afflictions that could be overcome if scientists were free to conduct their research.

Lee M. Silver reaches a similar conclusion, that while we wring our hands, Asia is poised to take the lead in the development of biotechnology. According to Silver, the opposing parties to biotechnology are largely ignorant of the history of biotechnology. Civilization was founded on biotechnology. The fruits, vegetables, and livestock that sustain modern society were not original gifts of nature, but the products of painstaking genetic alteration over thousands of years. Human beings have benefited from and lived with biotechnology from the beginning of civilization. Silver stops short of suggesting that there is nothing novel about the new biotechnology.

III

The novelty, of course, is the claim that biomedical science and biotechnology can remake human nature. Hava Tirosh-Samuelson notes that a distinction can be drawn between ancient and modern science. She shows through a thorough historical account of Jewish philosophy that ancient philosophy or science can be understood as contemplative and characterized by the search for knowledge of the material world. In contrast, modern science is concerned with making and doing, or in the words of Francis Bacon, the conquest of Nature for the relief of man's estate.

Arguably, the controversy surrounding biotechnology and biomedicine represents the larger doubts we harbor concerning the modern project and the conquest of Nature. The modern identity of scientific advancement and social progress is now open for debate.

It was only after the efforts of Bacon and Descartes that modern scientific inquiry and technical innovation came to be thought of as crucial for social progress. In terms of modern social contract theory the freedom to engage in scientific inquiry was granted on the grounds that it would alleviate man's estate. Modern science sought this harmony by directing itself to do and to make and to minister to the desires of the body, such that health and self-preservation became the highest human goods. But as both the biotechnical pessimists and optimists contend, the novelty of biomedical science and its technology is that we can go beyond nature's endowment and enhance our biological makeup. To make the point even more explicit, the claim of biotechnology is that we can take charge of our evolutionary future.

At the heart of the biotechnology debate is the distinction between enhancement and therapy. Originally, the distinction was intended to establish a criterion for judging between acceptable and unacceptable uses of biomedical technology, where therapy was ethically acceptable while enhancement was considered suspect. If therapy is intended to restore one to normal then we can see that it is different from going beyond normal, or what can be called enhancement. Kass argues that the distinction cannot be maintained. He points out that what is normal in some areas such as behavioral or psychic function is difficult to assess. Kass concludes that the distinction is a distraction from the more important question: What are the good uses of biotechnology?

Green agrees that the distinction between therapy and enhancement cannot strictly be maintained. He points out that much of what we consider to be therapy is indistinguishable from enhancement, such as the infant vaccination program that has taken the human immune system beyond what was considered normal. Accordingly, we might all be considered enhanced human beings in one way or another. Richard Sherlock in response to Green's essay in this volume notes that if biotechnology can transform human nature, then human nature and standards like normal have no meaning. He goes on to add that, likewise, enhancement is meaningless because one would need a fixed standard by which to determine whether the so called enhancement was an improvement or not.

Nevertheless what is most important for Green, and Bailey would certainly concur, is the middle ground between enhancement and therapy, disease prevention, which would include alterations of the human genome. Indeed this is the main thrust of the proponents of the biotechnological revolution, that human beings should be free to use their science to improve themselves and their children and free themselves from disease and affliction, rather than leave these things in the hands of nature, or chance. What is to be feared is not the technology, but heavy-handed government regulation and bureaucratic agencies that replace individual choice with the choices of faceless bureaucrats too willing to impose their personal predilections on others.

IV

Both the critics and the advocates of biomedical science and technology believe that its use will lead to a "post-human future," where human nature will be radically changed or even abolished. Critics fear that we will no longer have human nature as the natural ground of moral experience. To avoid this quandary, they advocate that we must look to the Bible and theology as a sound guide for judging the use of biotechnology. That religion shapes the character of a people and influences how they are to understand what is good, what is just, and their place in the universe cannot be denied.

Silver recognizes that while civilization was founded on the biotechnical advances in agriculture and animal husbandry, biotechnical and biomedical advances are limited by religion. Silver rightly observes that not all religions or "spiritual traditions" are the same. Christians, he asserts, view biotechnology as a violation of God's plan for each individual soul, while New Age secularists view agricultural biotechnology as a direct assault on Mother Nature. In contrast, Eastern religious traditions tend to view the soul as both eternal and self-evolving, not beholden to a divine plan. Eastern religions are therefore more open to embryonic research and genetically modified plants. It is to Asia, Silver maintains that we will look for the next breakthroughs in biomedical science and biotechnology.

Tirosh-Samuelson points out that Israel and many modern Jews strongly support biomedical science and technology. She contends that the Jewish philosophic tradition, represented by Philo of Alexandria, Moses Maimonides, Hans Jonas, and Joseph Dov Soloveitchik, offers a critical vantage point from which to evaluate the enthusiastic Jewish endorsement of biotechnology. The tendency of biotechnology to reduce humanness to the genetic makeup of the body is questioned by Jewish philosophy, which regards the creation of man in the image of God as that which make us distinctly human. She notes the tradition questions the modern project to control and manipulate nature, including the nature of man, which God has created. In this way, Tirosh-Samuelson argues, the Jewish philosophical tradition, in recognizing the distinction between the human and God, between the creature and the Creator, emphasizes the responsibility of humans to protect and nurture the created order.

Larry Arnhart disputes the claim that biotechnology will abolish human nature. That claim ignores how Darwinian evolution has shaped our bodies, our brains, and our desires in ways that resist technical manipulation. Biotechnology will be limited in its technical ends because our behavioral traits are rooted in the interplay between our genes and our individual life histories and unpredictable experiences, which offer unique developmental opportunities and together form highly adaptable unique brains. Viewed in this way, the technical manipulation of human nature to enhance desirable traits while avoiding undesirable side effects will be difficult, if not impossible. Much of what makes us unique human beings is not amendable to genetic manipulation, such as our individual life histories and experiences. We are not simply determined by our genes. Further, Arnhart does not agree with the claim that in order to judge the uses of biotechnology we must look to biblical wisdom. He argues that the Bible's moral teaching is sometimes unclear and sometimes unreliable, necessitating that we read the Bible in light of our natural moral sense. Indeed for Arnhart the natural moral sense and our natural desire for what is good, implanted in human beings by natural selection in evolutionary history, can be relied upon to guide us in our use of biotechnology. He reasons that if the good is desirable then morality is natural insofar as it satisfies the natural desires of our species. This is the basis of his claim that biotechnology will be limited in its moral ends because it will be directed by our natural desires and our moral sense of what is good and desirable. Moreover, Arnhart points out that both the proponents and critics of biotechnology are compelled to appeal to our natural moral sense. Proponents in making their claim that there is nothing to fear from the use of biotechnology appeal to our natural moral sense by saying that we should have confidence in the choices that individuals will make, that is, individuals will be guided by what is desirable. Critics appeal to our natural moral sense to make their argument for limiting the use of biotechnology and for defending its good uses.

While Arnhart is confident that human nature is here to stay, Sherlock is less certain. Sherlock argues that the natural moral sense needs theology to support it, because biotechnology can and does alter human nature. He suggests, in light of recent advances, that it is no longer possible to appeal to nature and human nature as the moral ground for regulating the use of biomedicine and biotechnology. Naturalism cannot be sustained in the age of biotechnology. Therefore, Sherlock concludes, that we must look to the transnaturalism of Christian theology in order to develop a sound notion of human dignity, to understand our place in the cosmos, and to articulate our obligations to our neighbors. In the end, Sherlock's theological position provides both a cautionary argument for the use of biomedicine and technology, as well as a case for the limited use of biotechnology grounded in Christian theology.

VI

While the debate concerning biomedical science and technology for the most part has focused upon the possibility that human nature can be transformed, we must also recognize that we are both human beings and citizens, that is, there are serious political and social questions at stake with the new technologies.

At times some have argued that biotechnology could bring about a truly equal society, particularly if the technology is allowed to level out nature's lottery. Green concedes that this is something he has considered. Certainly, one can see that the unequal use of biomedicine, in sports or in school, would offend our sense of fairness or possibly introduce biological distinctions within society. Bailey is appalled by this possibility—that there are those who view biotechnology as the scientific means to even out society. He argues, much like Arnhart, that this is an example of biodeterminism—you are what your genes determine. Bailey points out that this fails to understand how genes work. While we all have genetic dispositions, we may find that we choose to pursue different activities and choose to develop different faculties, so it is unlikely and dangerous to think that government could bring about a truly egalitarian society. We might also add that this notion of equality as equal faculties is not the same as the democratic understanding of equality of rights. Some have argued that through biotechnology we can choose the genetic makeup of future generations without their consent, that is, hold future generations hostage to current generations' notion of genetic perfection. The essays by Arnhart and Bailey suggest that this too does not take into account human freedom.

In the end, biotechnology and biomedicine have already impacted our society by forcing us to consider what it means to be a human being and what our obligations are as citizens to the sick, to the weak, and to nascent life. What should be clear to the reader of this volume is that the essays demonstrate that biotechnology occasions the consideration of the enduring question that has always intrigued human beings. What does it mean to be human?

ONE

Biotechnology and Our Human Future

Some General Reflections

LEON R. KASS

As nearly everyone appreciates, we live near the beginning of the golden age of biomedical science and technology. For the most part, we should be mightily glad that we do. We and our friends and loved ones are many times over the beneficiaries of its cures for diseases, prolongation of life, and amelioration of suffering, psychic and somatic. We should be deeply grateful for the gifts of human ingenuity and for the devoted efforts of scientists, physicians, and entrepreneurs who have used these gifts to make those benefits possible. And, mindful that modern biology is just entering puberty, we suspect that "ya' ain't seen nothin' yet."

Yet, notwithstanding these blessings, present and projected, we have also seen more than enough to make us concerned. For we recognize that the powers made possible by biomedical science can be used for nontherapeutic purposes, serving ends that range from the frivolous and disquieting to the offensive and pernicious. Biotechnologies are available as instruments of bioterrorism (for example, genetically engineered drug-resistant bacteria, or drugs that obliterate memory); as agents of social control (for example, drugs to tame rowdies and dissenters or fertility-blockers for welfare recipients); and as means of trying to improve or perfect our bodies and minds or those of our children (for example, genetically engineered "super-muscles," or drugs to improve memory or academic performance). Anticipating possible threats to our security, freedom, and even our very humanity, many people are increasingly worried about where biotechnology may be taking us. We are concerned not only about what others might do to us, but also about what we might do to ourselves. We are concerned that our society might be harmed and that we ourselves might be diminished, indeed, in ways that could undermine the highest and richest possibilities of human life.

In this essay I will consider only the last and most seductive of these disquieting prospects—the use of biotechnical powers to improve upon human nature or to pursue "perfection," both of body and of mind. I select this subject for several reasons. First, although it is the most neglected topic in public bioethics, it is I believe the deepest source of public anxiety about biotechnology and the human future, represented in concerns expressed about "man playing God" or about the arrival of the Brave New World or a "post-human future." Second, it raises the weightiest questions—questions about the ends and goals of the biomedical enterprise, the nature and meaning of human flourishing, and the intrinsic threat of dehumanization (or the promise of super-humanization). It therefore, third, compels attention to what it means to *be* a human being and to be active *as* a human being. Finally, it gets us beyond our narrow preoccupation with the "life issues" of abortion or embryo destruction, important though they are, to deal with what is genuinely novel in the biotechnical revolution: not the old, crude power to kill the creature made in God's image, but science-based sophisticated powers to remake him after our own fantasies.

What exactly are the powers that I am talking about? What sorts of ends are they likely to serve? How soon are they available? They are powers that affect the capacities and activities of the human body, powers that affect the capacities and activities of the mind or soul, and powers that affect the shape of the human life cycle, at both ends and in between. We already have powers to prevent fertility and to promote it; to initiate life in the laboratory; to screen our genes, both as adults and as embryos, and to select (or reject) nascent life based on genetic criteria; to insert new genes into various parts of the adult body and someday soon also into gametes and embryos; to enhance muscle performance and endurance; to replace body parts with natural or mechanical organs, and perhaps soon, to wire ourselves using computer chips implanted into the body and brain; to alter memory, mood, desire, temperament, and attention though psychoactive drugs; and to prolong not just the average but also the maximum human life expectancy. The technologies for altering our native capacities and activities are mainly those of genetic screening and genetic engineering; drugs, especially psychoactive ones; and the ability to replace body parts or to insert novel ones. The availability of some of these capacities, using these techniques, has been demonstrated only with animals; but others are already in use in humans.

It bears emphasis that these powers have not been developed for the purpose of producing perfect or post-human beings. To the contrary, they have been produced largely for the purpose of preventing and curing disease and of reversing disabilities. Even the bizarre prospects of machine-brain interaction and implanted nanotechnological devices start with therapeutic efforts to enable the blind to see and the deaf to hear. Yet the "dual use" aspects of most of these powers, encouraged by the ineradicable human urge toward "improvement" and the commercial interests that see market opportunities for such nontherapeutic uses, means that we must not be lulled to sleep by the fact that the originators of these powers were no friends to the Brave New World. Once here, techniques and powers can produce desires where none existed before, and things often go where no one ever intended—not least because each technological success in combating disease and disability seems only to increase popular demand for evermore-effective means of overcoming any and all remaining obstacles to satisfying our desires and working our wills.

How to organize our reflections? One should resist the temptation to begin with the new techniques or even with the capacities for intervention that they make possible. To do so runs the risk of losing the human import and significance of the undertakings. Better to begin with the human desires and goals that these powers and techniques are destined to serve, among them, the desires for better children, superior performance, ageless bodies, happy souls, and a more peaceful and cooperative society.[1] In this essay, I will concentrate mainly on the strictly personal goals of self-improvement and self-enhancement, and especially on those efforts to preserve and augment the vitality of the body and to increase the happiness of the soul. These goals are, arguably, the least controversial, the most continuous with the aims of modern medicine and psychiatry (better health, peace of mind), and the most attractive to most potential consumers—probably indeed to most of us. Indeed, these were the very goals, now at last in the realm of possibility, that animated the great founders of modern science, Francis Bacon and René Descartes: flawlessly healthy bodies, unconflicted and contented souls, and freedom from the infirmities of age, perhaps indefinitely.

Although our discussion here will not be driven by the biotechnologies themselves, it may be useful to keep in mind some of the technological approaches and innovations that, in varying degrees, can serve these purposes. For example, in pursuit of "ageless bodies," (1) we can replace worn-out parts, by means of organ transplantation or, in the future, by regenerative medicine where decayed tissues are replaced with new ones produced from stem cells; (2) we can improve upon normal and healthy parts, for example, via precise genetic modification of muscles, through injections of growth factor genes that keep the transformed muscles whole, vigorous, and free of age-related decline;[2] and (3) most radically, we can try to retard or stop the entire process of biological senescence. Especially noteworthy for this last possibility are recent discoveries in the genetics of aging that have shown how the *maximum* species lifespan of worms and flies can be increased two- and threefold by alterations in a *single* gene, a gene now known to be present also in mammals, including humans.

In pursuit of "happy souls," we can eliminate psychic distress, we can produce states of transient euphoria, and we can engineer more permanent conditions of good cheer, optimism, self-esteem, and contentment. Accordingly, please keep in mind the existence of drugs now available that, administered promptly at the time of memory formation, blunt markedly the painful emotional content of the newly formed memories of traumatic events (so-called memory blunting or erasure, a remedy being sought to prevent posttraumatic stress disorder). Keep in mind, second, the existence of euphoriants, like Ecstasy, the forerunner of Huxley's "soma," widely used on college campuses; and, finally, powerful yet seemingly safe antidepressants and mood brighteners like Prozac, wonderful for the treatment of major depression yet also capable in some people of utterly changing their outlook on life from that of Eeyore to that of Mary Poppins.

Problems of Description: The Distinction between Therapy and Enhancement

People who have tried to address our topic have usually approached it through a distinction between "therapy" and "enhancement": "therapy," the treatment of individuals with known diseases or disabilities; "enhancement," the directed uses of biotechnical power to alter, by direct intervention, not diseased processes but the "normal" workings of the human body and psyche. Those who introduced this distinction hoped by this means to distinguish between the acceptable and the dubious or unacceptable uses of biomedical technology: therapy is always ethically fine, enhancement is, at least *prima facie*, ethically suspect. Gene therapy for cystic fibrosis or Prozac for psychotic depression is fine; insertion of genes to enhance intelligence or steroids for Olympic athletes is not.[3]

This distinction is useful as a point of departure: restoring to normal does differ from going beyond the normal. But it proves finally inadequate to the moral analysis. Enhancement is, even as a term, highly problematic. Does it mean "more" or "better," and, if "better," by what standards? Can both improved memory and selective erasure of memory both be "enhancements"? If "enhancement" is defined in opposition to "therapy," one faces further difficulties with the definitions of "healthy" and "impaired," "normal" and "abnormal" (and hence, "super-normal"), especially in the area of "behavioral" or "psychic" functions and activities. Some psychiatric diagnoses are notoriously vague, and their boundaries indistinct: how does "social anxiety disorder" differ from shyness, "hyperactivity disorder" from spiritedness, "oppositional disorder" from the love of independence? Furthermore, in the many human qualities (for example, height or IQ) that distribute themselves "normally," does the average also function as a norm, or is the norm itself appropriately subject to alteration? Is it therapy to give growth hormone to a genetic dwarf but not to an equally short fellow who is just unhappy to be short? And if the short are brought up to the average, the average, now having become short, will have precedent for a claim to growth hormone injections. Needless arguments about whether or not something is or is not an "enhancement" get in the way of the proper questions: What are the good and bad uses of biotechnical power? What makes a use "good," or even (merely) "acceptable"? It does not follow from the fact that a drug is being taken solely to satisfy one's desires—for example, to sleep less or to concentrate more—that its use is objectionable. Conversely, certain interventions to restore natural functioning wholeness—for example, to enable postmenopausal women to bear children or sixty-year-old men to keep playing professional ice hockey—might well be dubious uses of biotechnical power.

This last observation points to the deepest reason why the distinction between healing and enhancing is finally insufficient, both in theory and in practice. For the human whole whose healing is sought or accomplished by biomedical therapy is finite and frail, medicine or no medicine. The healthy body declines and its parts wear out. The sound mind slows down and has trouble remembering things. The soul has aspirations beyond what even a healthy body can realize, and it becomes weary from frustration. Even at its fittest, the fatigable and limited human body rarely carries out flawlessly even the ordinary desires of the soul. Moreover, there is wide variation in the natural gifts with which each of us is endowed: some are born with perfect pitch, others are born tone-deaf; some have flypaper memories, others forget immediately what they have just learned. And as with talents, so too with the desires and temperaments: some crave immortal fame, others merely comfortable preservation. Some are sanguine, others phlegmatic, still others bilious or melancholic. When nature deals her cards, some receive only from the bottom of the deck.[4]

As a result of these infirmities, human beings have long dreamed of overcoming limitations of body and soul, in particular the limitations of bodily decay, psychic distress, and the frustration of human aspiration. Until now these dreams have been pure fantasies, and those who pursued them came crashing down in disaster. But the stupendous successes over the past century in all areas of technology, and especially in medicine, have revived the ancient dreams of human perfection. We major beneficiaries of modern medicine are less content than we are worried, less grateful for the gifts of longer life and better health and more anxious about losing what we have. Accordingly, we regard our remaining limitations with less equanimity, even to the point that dreams of getting rid of them can be turned into moral imperatives. For these reasons, thanks to biomedical technology, people will be increasingly tempted to pursue these dreams, at least to some extent: ageless and ever-vigorous bodies, happy (or at least not unhappy) souls, and excellent human achievement (with diminished effort or toil).

Why should anyone be bothered by these prospects? What could be wrong with efforts to improve upon human nature, to try, with the help of biomedical technology, to gain ageless bodies and happy souls? I begin with some familiar sources of concern.

Familiar Sources of Concern

Not surprisingly, the objections usually raised to the "beyond therapy" uses of biomedical technologies reflect the dominant values of modern America: health, equality, and liberty.

(1) *Health: issues of safety and bodily harm.* In our health-obsessed culture, the first reason given to worry about any new biological intervention is safety, and that is true also here. Athletes who take steroids will later suffer premature heart disease. College students who take Ecstasy will damage dopamine receptors in their basal ganglia and suffer early Parkinson's disease. To generalize: no biological agent used for purposes of self-perfection will be entirely safe. This is good conservative medical sense: anything powerful enough to enhance system A is likely to be powerful enough to harm system B. Yet many good things in life are filled with risks, and free people if properly informed may choose to run them, if they care enough about what is to be gained thereby. If the interventions are shown to be *highly* dangerous, many people will (later if not sooner) avoid them, and the Food and Drug Administration and/or tort liability will constrain many a legitimate purveyor. It surely makes sense, as an ethical matter, that one should not risk basic health pursuing a condition of "better than well." But, on the other hand, if the interventions work well and are indeed highly desired, people may freely accept, in trade-off, even considerable risk of later bodily harm. But in any case, the big issues have nothing to do with safety; as in the case of cloning children, the real questions concern not the safety of the procedures but what to think about the perfected powers, assuming that they may be safely used.

(2) *Equality: issues of unfairness and distributive justice.* An obvious objection to the use of personal enhancers by participants in competitive activities is that they give those who use them an unfair advantage: blood doping or steroids in athletes, stimulants in students taking the SATs. The objection has merit, but it does not reach to the heart of the matter. For even if *everyone* had *equal* access to brain implants or genetic improvement of muscle strength or mind-enhancing drugs, a deeper disquiet would still remain. Even were steroid or growth hormone use by athletes to be legalized, most athletes would be ashamed to be seen injecting themselves before coming to bat. Besides, not all activities of life are competitive: it would matter to me if she says she loves me only because she is high on "erotogenin," a new brain stimulant that mimics perfectly the feeling of falling in love. It matters to me

when I go to a seminar that the people with whom I am conversing are not drugged out of their right minds.

The distributive justice question is less easily set aside than the unfairness question, especially if there are systematic disparities between who will and who will not have access to the powers of biotechnical "improvement." The case can be made yet more powerful to the extent that we regard the expenditure of money and energy on such niceties as a misallocation of limited resources in a world in which the basic health needs of millions go unaddressed. It is embarrassing, to say the least, to discover that in 2002, for example, Americans spent one billion dollars on baldness, roughly ten times the amount spent worldwide for research on malaria. But, once again, the inequality of access does not remove our disquiet to the thing itself. And it is to say the least paradoxical, in discussions of the dehumanizing dangers of, say, eugenic choice, when people complain that the poor will be denied equal access to the danger: "The food is contaminated, but why are my portions so small?" Check it out: yes, Huxley's *Brave New World* runs on a deplorable and impermeably rigid class system, but would you want to live in that world if offered the chance to enjoy it as an alpha (one of the privileged caste)? Even an elite can be dehumanized, can dehumanize itself. The central matter is not equality of access, but the goodness or badness of the thing being offered.

(3) *Liberty: issues of freedom and coercion, overt and subtle.* This comes closer to the mark, especially with uses of biotechnical power exercised by some people upon other people, whether for social control—say in, the pacification of a classroom of Tom Sawyers—or for their own putative improvement—say, with genetic selection of the sex or sexual orientation of a child-to-be. This problem will of course be worse in tyrannical regimes. But there are always dangers of despotism within families, as parents already work their wills on their children with insufficient regard to a child's independence or long-term needs or the "freedom to be a child." To the extent that even partial control over genotype—say, to take a relatively innocent example, musician parents selecting a child with genes for perfect pitch—adds to existing social instruments of parental control and its risks of despotic rule, this matter will need to be attended to.

There are also more subtle limitations of freedom, say, through peer pressure. What is permitted and widely used may become mandatory. If most children are receiving memory enhancement or stimulant drugs to enable them to "get ahead," failure to provide them for your child might come to be seen as a form of child neglect. If all the defensive linemen are on steroids, you risk mayhem if you go against them chemically pure. And, a point subtler still, some critics complain that, as with cosmetic surgery, Botox, and breast implants, the enhancement technologies of the future will likely be used in slavish adherence to certain socially defined and merely fashionable notions of "excellence" or improvement, very likely shallow, almost certainly conformist.

This special kind of restriction of freedom—let's call it the problem of conformity or homogenization—is in fact quite serious. We are right to worry that the self-selected nontherapeutic uses of the new powers, especially where they become widespread, will be put in the service of the most common human desires, moving us toward still greater homogenization of human society—perhaps raising the floor but greatly lowering the ceiling of human possibility, and reducing the likelihood of genuine freedom, individuality, and greatness. Indeed, such homogenization may be the most important society-wide concern, if we consider the aggregated effects of the likely individual choices for biotechnical "self-improvement," each of which might be defended or at least not objected to on a case-by-case basis (the problem of what the economists call "negative externalities"). For example, it would be difficult to object to a personal choice for a life-extending technology that would extend the user's life by three healthy decades or a mood-brightened way of life that would make the individual more cheerful and untroubled by the world around him. Yet the aggregated social effects of such choices, widely made, could lead to a Tragedy of the Commons, where genuine and sought for satisfactions for individuals are nullified or worse, owing to the social consequences of granting them to everyone.[5] And, as Aldous Huxley strongly suggests in *Brave New World*, the use of biotechnical powers to produce contentment in accordance with democratic tastes threatens the character of human striving and diminishes the possibility of human excellence; perhaps the best thing to be hoped for is preservation of pockets of difference (as on the remote islands in *Brave New World*) where the desire for high achievement has not been entirely submerged in the culture of "the last man."

But, once again, important though this surely is as a social and political issue, it does not settle the question regarding individuals. What if anything can we say to justify our disquiet over the individual uses of performance-enhancing genetic engineering or mood-brightening drugs for other than medical necessity? For even the safe, equally available, non-coerced and non-faddish uses of these technologies for "self-improvement" raise ethical questions, questions that are at the heart of the matter: the disquiet must have something to do with the essence of the activity itself, the use of technological means to intervene into the human body and mind not to ameliorate disease but to change and improve their normal workings. Why, if at all, are we bothered by the voluntary *self*-administration of agents that would change our bodies or alter our minds? What is disquieting about our freely chosen attempts to improve upon human nature, or even our own particular instance of it?

It will be difficult, I acknowledge at the outset, to put this disquiet into words. Initial repugnances are hard to translate into sound moral arguments. We are probably repelled by the idea of drugs that would erase memories or change personalities, or interventions that might enable seventy-year-olds to bear children or play professional sports, or, to engage in some wilder imaginings, the prospect of mechanical implants that would enable men to nurse infants or computer/body hookups that would enable us to download the *Oxford English Dictionary*. But is there wisdom in this repugnance? Taken one person at a time, with a properly prepared set of conditions and qualifications, it is going to be hard to say what is wrong with any biotechnical intervention that could give us (more) ageless bodies or superior performances or could make it possible for us to have happier souls.

If there are essential reasons to be concerned about these activities and where they may lead us, we sense that they must have something to do with challenges to what is naturally human, to what is humanly dignified, or to attitudes that show proper respect for what is naturally and dignifiedly human. In reverse order, I will make three arguments, one on each of these three themes: respect for "the naturally given," threatened by hubris; the dignity of unadulterated human activity, threatened by "unnatural" means; and the nature of full human flourishing, threatened by spurious, partial, or shallow substitutes.

Hubris or Humility? Respect for "the Given"

A common man-on-the-street reaction to these prospects is the complaint of "men playing God." If properly unpacked, this worry is in fact shared by people holding various theological beliefs and by people holding none at all. Sometimes the charge means the sheer prideful presumption of trying to alter what God has ordained or nature has produced, or what should, for whatever reason, not be fiddled with. Sometimes the charge means not so much usurping godlike powers, but doing so in the absence of godlike knowledge: the mere playing at being God, the hubris of acting with insufficient wisdom.

Over the past few decades, environmentalists, forcefully making the case for respecting Mother Nature, have urged upon us a "precautionary principle" regarding our interventions into the natural world. Go slowly, they say, you can ruin everything. The point is certainly well taken in the present context. The human body and mind, highly complex and delicately balanced as a result of eons of gradual and exacting evolution, are almost certainly at risk from any ill-considered attempt at "improvement." There is not only the matter of unintended consequences, a concern present even with interventions aimed at therapy. There is also the matter of uncertain goals and absent natural standards, once one proceeds "beyond therapy." When a physician intervenes therapeutically to correct some deficiency or deviation from a patient's natural wholeness, he acts as a servant to the goal of health and as an assistant to nature's own powers of self-healing, themselves wondrous products of evolutionary selection. But when a bioengineer intervenes to "improve upon nature," he stands not as nature's servant but as her aspiring master, guided by nothing but his own will and serving ends of his own devising. It is far from clear that our delicately integrated natural bodily powers will take kindly to such impositions, however desirable the sought-for change may seem to the overconfident intervener. And there is the further question of the goodness of the goals being sought, a matter to which I will return.

One revealing way to formulate the problem of hubris is what Michael Sandel has called the temptation to "hyper-agency," a Promethean aspiration to remake nature, including human nature, to serve our purposes and to satisfy our desires. This attitude is to be faulted not only because it can lead to bad, unintended consequences; more fundamentally, it also represents a false understanding of, and an improper disposition toward, the naturally given world. The root of the difficulty, according to Sandel, seems to be both cognitive and moral: the failure properly to appreciate and respect the "giftedness" of the world.

To acknowledge the giftedness of life is to recognize that our talents and powers are not wholly our own doing, nor even fully ours, despite the efforts we expend to develop and to exercise them. It is also to recognize that not everything in the world is open to any use we may desire or devise. An appreciation of the giftedness of life constrains the Promethean project and conduces to a certain humility. It is, in part, a religious sensibility. But its resonance reaches beyond religion.[6]

The point is well taken, as far as it goes, for the matter of our attitude toward nature is surely crucial. Human beings have long manifested both wondering appreciation for nature's beauty and grandeur and reverent awe before nature's sublime and mysterious power. From the elegance of an orchid to the splendor of the Grand Canyon, from the magnificence of embryological development to the miracle of sight or consciousness, the works of nature can still inspire in most human beings an attitude of respect, even in this age of technology. Nonetheless, the absence of a respectful attitude is today a growing problem in many quarters of our high-tech world. It is worrisome when people act toward, or even talk about, our bodies and minds—or human nature itself—as if they were mere raw material to be molded according to human will. It is worrisome when people speak as if they were wise enough to redesign human beings, improve the human brain, or reshape the human life cycle. In the face of such hubristic temptations, appreciating that the given world—including our natural powers to alter it—is not of our own making could induce a welcome attitude of modesty, restraint, and humility. Such an attitude is surely recommended for anyone inclined to modify human beings or human nature for purposes beyond therapy.

Yet the respectful attitude toward the "given," while both necessary and desirable as a restraint, is not by itself sufficient as a guide. The "giftedness of nature" also includes smallpox and malaria, cancer and Alzheimer's disease, decline and decay. Moreover, nature is not equally generous with her gifts, even to human beings, the most gifted of her creatures. Modesty born of gratitude for the world's "givenness" may enable us to recognize that not everything in the world is open to any use we may desire or devise, but it will not *by itself* teach us *which* things can be tinkered with and *which* should be left inviolate. Respect for the "giftedness" of things cannot tell us which gifts are to be accepted as is, which are to be improved through use or training, which are to be housebroken through self-command or medication, and which opposed like the plague.

The word "given" has two relevant meanings, the second of which Sandel's account omits: "given," meaning "bestowed as a gift," and "given" (as in mathematical proofs), something "granted," definitely fixed and specified. Most of the given bestowals of nature have their given species-specified *natures*: they are each and all of a given *sort*. Cockroaches and humans are equally bestowed but differently natured. To turn a man into a cockroach would be dehumanizing. To try to turn a man into more than a man might be so as well. To avoid this, we need more than generalized appreciation for nature's gifts. We need a particular regard and respect for the special gift that is our own given nature.

In short, only if there is a human "givenness," or a given "humanness," that is also *good and worth respecting*, either as we find it or as *it could be perfected without ceasing to be itself*, will the "given" serve as a *positive* guide for choosing what to alter and what to leave alone. Only if there is something precious *in our given human nature*—beyond the mere fact of its giftedness—can what is given guide us in resisting efforts that would degrade it. When it comes to human biotechnical engineering beyond therapy, only if there is something inherently good or dignified about, say, natural procreation, the human life cycle (with its rhythm of rise and fall), and human erotic longing and striving; only if there is something inherently good or dignified about the ways in which we engage the world as spectators and appreciators, as teachers and learners, leaders and followers, agents and makers, lovers and friends, parents and children, citizens and worshippers, and as seekers of our own special excellence and flourishing in whatever arena to which we are called, only then can we begin to see why those aspects of our nature need to be defended against our deliberate redesign.

We must move, therefore, from the danger of hubris in the powerful designer to the danger of degradation in the designed, considering how any proposed improvements might impinge upon the nature of the one being improved. With the question of human nature and human dignity in mind, we move to questions of means and ends.

"Unnatural" Means: The Dignity of Human Activity

Until only yesterday, teaching and learning or practice and training exhausted the alternatives for acquiring human excellence, perfecting our natural gift through our own efforts. But perhaps no longer: biotechnology may be able to do nature one better, even to the point of requiring no teaching and less training or practice to permit an improved nature to shine forth. The insertion of the growth factor gene into the muscles of rats and mice bulks them up and keeps them strong and sound without the need for nearly as much exertion. Drugs to improve memory, alertness, and amiability could greatly relieve the need for exertion to acquire these powers, leaving time and effort for better things. What, if anything, is disquieting about such means of gaining improvement?

The problem cannot be that they are artificial, unnatural, in the sense of having man-made *origins*. Beginning with the needle and the fig leaf, man has from the start been the animal that uses art to improve his lot. By our very nature, we are constantly looking for ways to better our lives through artful means and devices, for we humans are creatures with what Rousseau called "perfectibility." Ordinary medicine makes extensive use of artificial means, from drugs to surgery to mechanical implants, in order to treat disease. If the use of artificial means is absolutely welcome in the activity of healing, it cannot be their unnaturalness alone that makes us uneasy when they are used to make people "better than well."

Still, in those areas of human life in which excellence has until now been achieved only by discipline and effort, the attainment of those achievements by means of drugs, genetic engineering, or implanted devices looks to many people to be "cheating" or "cheap." Many people believe that each person should work hard for his achievements. Even if one prefers the grace of the natural athlete or the quickness of the natural mathematician—people whose performance deceptively appears to be effortless—we admire also those who overcome obstacles and struggle to try to achieve the excellence of the former. This matter of character—the merit of disciplined and dedicated striving—though *not* the deepest basis of one's objection to biotechnological shortcuts, is surely pertinent. For character is not only the source of our deeds, but also their product. Healthy children whose disruptive behavior is "remedied" by pacifying drugs rather than by their own efforts are not learning self-control; if anything, they are learning to think it unnecessary. People who take pills to block out from memory the painful or hateful aspects of a new experience will not learn how to deal with suffering or sorrow. A drug to induce fearlessness does not produce courage.

Yet things are not so simple. Some biotechnical interventions may assist in the pursuit of excellence without cheapening its attainment. And many of life's excellences have nothing to do with competition or adversity. Drugs to decrease drowsiness or increase alertness, sharpen memory, or reduce distraction may actually help people to pursue their natural goals of learning or painting or performing their civic duty. Drugs to steady the hand of a neurosurgeon or to prevent sweaty palms in a concert pianist cannot be regarded as "cheating," for they are not the source of the excellent activity or achievement. And, for people dealt a meager hand in the dispensing of nature's gifts, it should not be called cheating or cheap if biotechnology could assist them in becoming better equipped—whether in body or in mind.

Nevertheless, there is a sense here where the issue of "naturalness" of means matters. It lies not in the fact that the assisting drugs and devices are artifacts, but in the danger that they will violate or distort human agency and undermine the dignity of the naturally human way of being-at-work in the world. Here, in my opinion, is one of the more profound ways in which the use of at least some of these biotechnological means of seeking perfection—those that work on the brain—come under grave suspicion. In most of our ordinary efforts at self-improvement, whether by practice or training or study, we sense the relation between our doings and the resulting improvement, between the means used and the end sought. There is an experiential and intelligible connection between means and ends; we can see how confronting fearful things might eventually enable us to cope with our fears. We can see how curbing our appetites produces self-command. The capacity to be improved is improved by using it; the deed to be perfected is perfected by doing it.

In contrast, biomedical interventions act directly on the human body and mind to bring about their effects on a subject who is not merely passive but who plays no role at all. He can at best *feel* their effects *without understanding their meaning in human terms.*[7] Thus, a drug that brightened our mood would alter us without our understanding how and why it did so—whereas a mood brightened as a fitting response to the arrival of a loved one or an achievement in one's work is perfectly, because humanly, intelligible. And not only would this be true about our states of mind. *All* of our encounters with the world, both natural and interpersonal, would be mediated, filtered, and altered. Human experience under biological intervention becomes increasingly mediated by unintelligible forces and vehicles, separated from the human significance of the activities so altered. The relations between the knowing subject and his activities, and between his activities and their fulfillments and pleasures, are disrupted. The importance of human effort in human achievement is here properly acknowledged: the point is not the exertions of good character against hardship, but rather the humanity of an alert and self-experiencing agent making his deeds flow intentionally from his willing, knowing, and embodied soul.[8]

To be sure, an increasing portion of modern life is mediated life: the way we encounter space and time, the way we "reach out and touch somebody" via the telephone or internet. And one can make a case that there are changes in our souls and dehumanizing losses that accompany the great triumphs of modern technology. Life becomes easier, but, at the same time, it becomes less "real" and less immediate, as all our encounters with each other and the world are increasingly filtered through the distorting lenses of our clever devices and crude images. But so long as these technologies do not write themselves directly into our bodies and minds, we are in principle able to see them working on us, and free (again, in principle) to walk away from their use (albeit sometimes only with great effort). Once they work on us in ways beyond our ken, we are, as it were, passive subjects of what might as well be "magic." We become, in a sense, more and more like artifacts, creatures of our chemists and bioengineers.

The same point can perhaps be made about enhanced achievements as about altered mental states: to the extent that an achievement is the result of some extraneous intervention, it is detachable from the agent whose achievement it purports to be. That I can use a calculator to do my arithmetic does not make me a knower of arithmetic; if computer chips in my brain were to "download" a textbook of physics, would that make me a knower of physics? Admittedly, this is not always an obvious point to make: if I make myself more alert through Ritalin or coffee, or if drugs can make up for lack of sleep, I may be able to learn more using my unimpeded native powers and in ways to which I can existentially attest that it is *I* who is doing the learning. Still, if human flourishing means not just the accumulation of external achievements and a full curriculum vitae but a lifelong *being-at-work* exercising one's *human* powers *well* and without great impediment, our genuine happiness requires that there be little gap, if any, between the dancer and the dance.

Like dancing, most of life's activities are, to repeat, noncompetitive; most of the best of them—loving and working and savoring and learning—are self-fulfilling beyond the need for praise and blame or any other external reward. Indeed, in these activities, there is at best no goal beyond the activity itself. Such for-itself human-being-at-work-in-the-world, unimpeded and wholehearted, is what we are eager to preserve against dilution and distortion.

In a word: one major trouble with biotechnical (especially mental) "improvers" is that they produce changes in us by disrupting the normal character of human being-at-work-in-the-world, what Aristotle called *energeia psyches*, activity of soul, which, when fine and full constitutes human flourishing. With biotechnical interventions that skip the realm of intelligible meaning, we cannot really own the transformations nor experience them *as genuinely ours*. And we cannot know whether the resulting conditions and activities of our bodies and our minds are, in the fullest sense, our own *as human*.

Partial Ends, Full Flourishing

In taking up first the matter of questionable means for pursuing excellence and happiness, we have put the cart before the horse: we have neglected to speak about the goals. The issue of good and bad means must yield to the question about good and bad ends.

What do we think about the goals of ageless bodies and happy souls? Would their attainment in fact improve or perfect our lives *as* human beings? These are very big questions, too long to be properly treated here. But the following considerations seem to merit attention.

The case for ageless bodies seems at first glance to look pretty good. The prevention of decay, decline, and disability, the avoidance of blindness, deafness, and debility, the elimination of feebleness, frailty, and fatigue all seem to be conducive to living fully as a human being at the top of one's powers—of having a good "quality of life" from beginning to end. We have come to expect organ transplantation for our worn-out parts. We will surely welcome stem-cell-based therapies for regenerative medicine, reversing by replacement the damaged tissues of Parkinson's disease, spinal cord injury, and many other degenerative disorders. It is hard to see any objection to obtaining in our youth a genetic enhancement of all of our muscles that would not only prevent the muscular feebleness of old age but would empower us to do any physical task with much greater strength and facility throughout our lives. And, should aging research deliver on its promise of adding not only extra life to years but also extra years to life, who would refuse it? Even if you might consider turning down an ageless body for yourself, would you not want it for your beloved? Why should she not remain to you as she was back then when she first stole your heart? Why should her body suffer the ravages of time?

To say no to this offer seems perverse, but I would suggest that it is not. Indeed, the deepest human goods may be ours only because we live our lives in aging bodies, made mindful of our living in time and inseparable from the natural life cycle through which each generation gives way to the one that follows it. Yet because this argument is so counterintuitive, we need to begin not with the individual choice for an ageless body, but to look first at what the individual's life might look like in a world in which everyone made the same choice. We need to make the choice universal, and see the meaning of that choice in the mirror of its becoming the norm.

What if everybody lived life to the hilt, even as they approached an ever-receding age of death in a body that looked and functioned—let's not be too greedy—like that of a thirty year old? Would it be good if each and all of us lived like light bulbs, burning as brightly from beginning to end, but then popping off without warning, leaving those around us suddenly in the dark? Or is it perhaps better that there be a shape to life, everything in its due season, the shape also written, as it were, into the wrinkles of our bodies that live it? What would the relations between the generations be like if there never came a point at which a son surpassed his father in strength or vigor? What incentive would there be for the old to make way for the young, if the old slowed down but little and had no reason to think of retiring—if Michael could play until he were not forty but eighty or if most members of Congress could serve for more than sixty years? And might not even a moderate prolongation of life span with vigor lead to a prolongation in the young of functional immaturity—of the sort that has arguably already accompanied the great increase in average life expectancy experienced in the past century? One cannot think of enhancing the vitality of the old without retarding the maturation of the young.

Going against both common intuition and my own love of life, I have tried elsewhere to make a rational case for the blessings of mortality. In an essay entitled "*L'Chaim* and Its Limits: Why Not Immortality?"[9] I suggest that living self-consciously with our finitude is the condition of the possibility of many of the best things in human life: engagement, seriousness, a taste for beauty, the possibility of virtue, the ties born of procreation, the quest for meaning. Though the arguments are made against the case for immortality, they have weight also against even more modest prolongations of the maximum life span, especially in good health, that would permit us to live as if there were always tomorrow. For it is, I submit, only our ability to number our days that enables us to make them count.

Although human beings are understandably reluctant to grow old and die, and although many religions offer us the promise of a better life hereafter, I contend that the human desire for immortality is in fact a desire not so much for deathlessness as for something transcendent and perfect. It is therefore not a desire that the biomedical conquest of aging or the possession of an ageless body can satisfy. No amount of prolonging earthly life—not even a limitless period of "more of the same"—will answer our deepest longings, namely, longings for wholeness, wisdom, goodness, or godliness. Indeed, our relentless pursuit of perfect bodies and further life extension will deflect us—may indeed already be deflecting us—from realizing more fully the aspirations to which our lives naturally point, and from living well rather than merely staying alive.

A preoccupation with personal agelessness is finally incompatible with accepting the need for procreation and human renewal. Both for individuals and for a whole society, to covet a prolonged life span for ourselves is both a sign and a cause of our failure to open ourselves to procreation and to any higher purpose. It is probably no accident that it is a generation whose intelligentsia proclaims the death of God and the meaninglessness of life that embarks on life's indefinite prolongation and that seeks to cure the emptiness of life by extending it forever. For the desire to prolong youthfulness is not only a childish desire to eat one's life and keep it; it is also an expression of a childish and narcissistic wish incompatible with devotion to posterity. It seeks an endless present, isolated from anything truly eternal and severed from any true continuity with past and future. It is in principle hostile to children, because children, those who come after, are those who will take one's place;

they are life's answer to mortality, and their presence in one's house is a constant reminder that one no longer belongs to the frontier generation. One cannot pursue agelessness for oneself and remain faithful to the spirit and meaning of perpetuation.

Those who think that having an ageless body would solve the problems of growing old ignore the psychological effects simply of the passage of time—of experiencing and learning about the way things are. After a while, no matter how healthy we are, no matter how respected and well placed we are socially, most of us cease to look upon the world with fresh eyes. Little surprises us, nothing shocks us, righteous indignation at injustice dies out. We have seen it all already, seen it all. We have often been deceived; we have made many mistakes of our own. Many of us become small–souled, having been humbled not by bodily decline or the loss of loved ones but by life itself. So our ambition also begins to flag, or at least our noblest ambitions. As we grow older, Aristotle already noted, we "aspire to nothing great and exalted and crave the mere necessities and comforts of existence." At some point, most of us turn and say to our intimates, "Is this all there is?" We settle, we accept our situation—if we are lucky enough to be able to accept it. In many ways, perhaps in the most profound ways, most of us go to sleep long before our deaths—and we might even do so earlier in life if awareness of our finitude no longer spurred us to make something of ourselves.

Finally, a world devoted to ageless bodies paradoxically would not lead us to appreciate life or celebrate the health we have. On the contrary, as we have seen in recent decades, it would likely be a world increasingly dominated by anxiety over health and fear of death, intolerant of all remaining infirmity and disability and absolutely outraged by the necessity of dying, now that each of us is, like Achilles, seemingly but a heel short of immortality.

Assume for the sake of the argument that some of these consequences would follow from a world of greatly increased longevity and vigor: would it be simply good to have an ageless body? Is there not wisdom and goodness in the natural human life cycle, roughly three multiples of a generation: a time of coming of age; a time of flourishing, ruling, and replacing of self; and a time of savoring and understanding, but still sufficiently and intimately linked to one's descendants to care about their future and to take a guiding, supporting, and cheering role?

And what about pharmacologically assisted happy souls? Painful and shameful memories are disquieting; guilty consciences disturb sleep; low self-esteem, melancholy and world-weariness besmirch the waking hours. Why not memory blockers for the former, mood brighteners for the latter, and a good euphoriant—without risks of hangovers or cirrhosis—when celebratory occasions fail to be jolly? For let us be clear: if it is imbalances of neurotransmitters—a modern equivalent of the medieval doctrine of the four humors—that are responsible for our state of soul, it would be sheer priggishness to refuse the help of pharmacology for our happiness, when we accept it guiltlessly to correct for an absence of insulin or thyroid hormone.

The problem with pursuing a happy soul differs from the problem with pursuing an ageless body. An ageless body, I have argued, is a goal incompatible with preserving our full humanity. Being happy, however, would seem to be precisely a proper, even *the* proper, goal of human life. Still, seeking happiness through pharmacology is dubious on two grounds, each having to do with the shrunken view of "happiness" that informs such a quest and the limited (and limiting) sort of happiness that is obtainable with the aid of drugs. Regarding the removal of psychic troubles, it turns out that some suffering and unhappiness are probably good for us; regarding the creation of psychic satisfactions, it turns out that the mere fragrance of happiness gets mistaken for its real flowering.

Notwithstanding the reality of serious mental illness and the urgent need to treat it (with drugs, of course, if necessary), a little reflection makes clear that there is something misguided about the pursuit of *utter* psychic tranquility or the attempt to eliminate shame, guilt, and all painful memories. Traumatic memories, shame, and guilt, are, it is true, psychic pains. In extreme doses, they can be crippling. Yet short of the extreme, they can also be helpful and fitting. They are appropriate responses to horror, disgraceful conduct, and sin, and as such help us to avoid or fight against them in the future. Witnessing a murder *should* be remembered as horrible; doing a beastly deed *should* trouble one's soul. Righteous indignation at injustice depends on being able to feel injustice's sting. An untroubled soul in a troubling world is a shrunken human being. Moreover, to deprive oneself of one's true memories—in their truthfulness also of feeling—is to deprive oneself of one's own life and identity.

The positive feeling-states of soul (especially those inducible by drugs), though perhaps accompaniments of human flourishing, are not its essence. Ersatz pleasure or feelings of self-esteem are not the real McCoy. They are at most but shadows divorced from *the underlying human activities that are the essence of human flourishing*. Not even the most doctrinaire hedonist wants to have the pleasure that comes from playing baseball without swinging the bat or catching the ball. No music lover would be satisfied with getting from a pill the pleasure of listening to Mozart without ever hearing the music. Most people want both to feel good and to feel good about themselves, but only as a result of being good and doing good.

At the same time, there appears to be a connection between the possibility of feeling deep unhappiness and the prospects for genuine happiness. If one cannot grieve, one has not loved.[10] And to be capable of aspiration, one must know and feel lack. As Wallace Stevens put it, "Not to have is the beginning of desire." In short, if human fulfillment depends on our being creatures of need and finitude and therewith of longing and attachment, there may be a double-barreled error in the pursuit of ageless bodies and factitiously happy souls: far from bringing us what we really need, pursuing these partial goods could deprive us of the urge and energy to seek a richer and more genuine flourishing.

It is, indeed, the peculiar gift of our humanity to recognize the linkage between our unavoidable finitude and our higher possibilities. As Plato's Socrates observed long ago (in the *Symposium*), the heart of the human soul is *eros*, an animating power born of lack but pointed upward. At bottom, human *eros* is the fruit of the peculiar conjunction of and competition between two competing aspirations conjoined in a single living body, both tied to our finitude: the impulse to self-preservation and the urge to reproduce. The first is a self-regarding concern for our own personal permanence and satisfaction; the second is a self-denying aspiration for something that transcends our own finite existence, and for the sake of which we spend and even give our lives. Other animals, of course, live with these twin and opposing drives. But only the human animal is conscious of their existence and is driven to devise a life based in part on the tension between them. In consequence, only the human animal has explicit and conscious longings for something higher, something whole, something eternal, longings that we would not have were we not the conjunction of this bodily "doubleness," elevated and directed upward through conscious self-awareness. Nothing humanly fine, let alone great, will come out of a society that has crushed the source of human aspiration, the germ of which is to be found in the meaning of the sexually complementary "two" that seek unity and wholeness, and willingly devote themselves to the well-being of their offspring. Nothing humanly fine, let alone great, will come out of a society that is willing to sacrifice all other goods to keep the present generation alive and intact. Nothing humanly fine, let alone great, will come from the desire to pursue bodily immortality or pharmacological happiness for ourselves.

Looking into the future at goals pursuable with the aid of new biotechnologies, we can turn a reflective glance at our present human condition and the prospects now available to us to live a flourishing human life. For us today, assuming that we are blessed with good health and a sound mind, a flourishing human life is not a life lived with an ageless body or an untroubled soul, but rather a life lived in rhythmed time, mindful of time's limits, appreciative of each season and filled first of all with those intimate human relations that are ours only because we are born, age, replace ourselves, decline, and die—and know it. It is a life of aspiration, made possible by and borne of experienced lack, of the disproportion between the transcendent longings of the soul and the limited capacities of our bodies and minds. It is a life that stretches toward some fulfillment to which our natural human soul has been oriented, and, unless we extirpate the source, will always be oriented. It is a life not of better genes and enhancing chemicals but of love and friendship, song and dance, speech and deed, working and learning, revering and worshipping.

If this is true, then the pursuit of an ageless body may prove finally to be a distraction and a deformation. The pursuit of an untroubled and self-satisfied soul may prove to be deadly to desire, if finitude recognized spurs aspiration and fine aspiration acted upon *is itself* the core of happiness. Not the agelessness of the body, not the contentment of the soul, and not even the list of external achievements and accomplishments of life, but the engaged and energetic being-at-work of what nature uniquely gave to us is what we need to treasure and defend. All other "perfections" may turn out to be at best but passing illusions, at worst a Faustian bargain that could cost us our full and flourishing humanity.

Notes

1. The 2003 report of the President's Council on Bioethics on this topic, *Beyond Therapy: Biotechnology and the Pursuit of Happiness*, is organized around the first four of these themes. The complete text is available online at www.bioethics.gov or in two commercially reprinted editions published by ReganBooks (HarperCollins) and the Dana Press. This essay, in many places, draws on the council's report.

2. These powers, already used to produce "mighty mouse" and "super rat," will soon be available for treatment of muscular dystrophy and muscle weakness in the elderly. They will also be of interest to football and wrestling coaches and to the hordes of people who spend several hours daily pumping iron or sculpting their bodies.

3. Health-care providers and insurance companies have for now bought into this distinction, paying for treatment of disease and disability but not for enhancements.

4. Curiously—but, on reflection, not surprisingly—it is often the most gifted and ambitious who most resent their limitations: Achilles was willing to destroy everything around him, so little could he stomach that he was but a heel short of immortality.

5. I myself will later argue such a case with respect to the goal of increasing longevity with ageless bodies.

6. See his "What's Wrong with Enhancement?" a working paper prepared for the President's Council on Bioethics (http://www.bioethics.gov/background/sandelpaper.html). See also his "The Case against Perfection" in the April 2004 issue of *The Atlantic*.

7. So do alcohol and caffeine and nicotine, though, it should be pointed out, we use these agents not as pure chemicals but in forms and social contexts that, arguably, give them a meaning different from what they would have were we to take them as pills. Besides, our acceptance of these "drugs" cannot, without extensive further argument, serve as precedent or moral justification for accepting newer psychoactive enhancers. On the contrary, concerns about the newer possibilities may rightly serve to clarify and intensify our misgivings about these age-old "uppers" and "downers."

8. The lack of "authenticity" sometimes complained of in these discussions is not so much a matter of "playing false" or of not expressing one's "true self," as it is a departure from "genuine," unmediated, and (in principle) self-transparent human activity.

9. It appears as the penultimate chapter of my book, *Life, Liberty and the Defense of Dignity: The Challenge for Bioethics* (Encounter Books, 2002). The discussion in the next few paragraphs borrows heavily from that essay.

10. As C. S. Lewis observed profoundly, speaking about his grief, "The pain I feel now is the happiness I had before. That's the deal."

TWO

Who's Afraid of Posthumanity?

A Look at the Growing Left/Right Alliance in Opposition to Biotechnological Progress

RONALD BAILEY

Politics makes strange bedfellows, it is said, but I want to suggest to you that biotechnological progress is making some really strange bedfellows. Opposition to biotechnological progress is pulling together an extraordinary coalition of left-wing and right-wing fundamentalists into what longtime anti-biotechnology activist Jeremy Rifkin calls "Fusion BioPolitics." As Rifkin writes, "The biotech era will bring with it a very different constellation of political visions and social forces just as the industrial era did. The current debate over cloning human embryos and stem cell research is already loosening the old alliances and categories. It is just the beginning of the new biopolitics."[1] Unfortunately, events are proving him all too right.

These right-wing and left-wing fundamentalists are being drawn together by their fear of what they call "posthumanity." They fear that advances in biotechnology will soon permit parents and doctors to gain too much control over crucial aspects of human health and human flourishing. For example, biotech enhancements, either genetic or pharmacologic, might dramatically extend life spans, significantly boost intellectual capacities, and alter reproductive and family relations. By going beyond the current limits imposed on us by mortality, aging, sickness, and reproduction, biotechnology will produce posthumans who, paradoxically in the view of these left-wing and right-wing bioconservative fundamentalists, will be both greater and lesser than ourselves.

So time to name names: Who are those prominent intellectuals who are afraid of an allegedly posthuman future? On the Left, activists crowd the streets from Seattle to Johannesburg, protesting the development of genetically enhanced crops. Perennial anti-biotech activist Jeremy Rifkin asserts that biotech advances violate "the boundaries between the sacred and the profane"[2] and demands "a strict global moratorium, no release of GMOs (genetically modified organisms) into the environment."[3] Rifkin is now ably assisted by his disciple, environmentalist writer Bill McKibben, who despairs, "I think genetically engineering our children will be the worst choice human beings ever make."[4] George Annas, a lawyer and bioethicist at Boston University, is proposing a global ban on reproductive cloning and all interventions in the human germline including those aimed at curing genetic diseases.[5] Marcy Darnovsky from the Center for Genetics and Society and Tom Athanasiou from EcoEquity claim, "The techno-eugenic vision urges us, in case we still harbor vague dreams of human equality and solidarity, to get over them." They fear that biotechnology will "allow inequality to be inscribed in the human genome."[6] Richard Hayes, director of the Center for Genetics and Society, declares that the movement opposing human genetic engineering "will need to be of the same intensity, scope, and scale as the great movements of the past century that struggled on behalf of working people, anti-colonialism, civil rights, peace and justice, women's equality, and environmental protection."[7] And bioethicist Daniel Callahan, one of the cofounders of the Hastings Center, which is arguably the world's first bioethics think tank, argues that using genetic engineering and other biomedical technologies to extend human life beyond three score and ten is ethically illegitimate.[8]

On the Right, the most prominent bioconservative is Leon Kass from the University of Chicago, former chairman of President Bush's Council on Bioethics. Kass fiercely opposed in vitro fertilization in the 1970s and now opposes any type of human cloning and all future genetic intervention in the human germline. Kass warns that biotechnologists who say that they "merely want to improve our capacity to resist and prevent diseases, diminish our propensities for pain and suffering, decrease the likelihood of death" are deceiving themselves and us. Behind these modest goals, he says, actually lies a utopian project to achieve "nothing less than a painless, suffering-free, and, finally, immortal existence."[9] Human reproductive cloning must be banned based on what Kass calls the "wisdom of repugnance."[10] Francis Fukuyama approvingly notes,

> In Europe, the environmental movement is more firmly opposed to biotechnology than is its counterpart in the United States and has managed to stop the proliferation of genetically modified foods there dead in its tracks. But genetically modified organisms are ultimately only an opening shot in a longer revolution and far less consequential than the human biotechnologies now coming on line.[11]

And Adam Wolfson, former editor of the neoconservative journal *The Public Interest*, warns against the hubristic temptations offered by the biotech revolution: "So let's not fool ourselves: A sentiment less generous than education of the young drives the ambition to engineer smarter, cleverer beings. It is the desire for an even more complete mastery over nature."[12]

So what are those who oppose human genetic engineering so afraid of? First, keep in mind that enabling parents to genetically enhance their children is not going to be as easy as some of us hope, nor will it happen as soon as we might wish, but nearly everyone, especially perhaps those who fear the possibility most, agrees that one day it will be possible.

So where do we stand now? The first test tube baby, Louise Joy Brown, was born in 1978 thanks to Drs. Edwards and Steptoe. Since then more than one million parents have used in vitro fertilization to have families. Dr. Edwards recently estimated that as many as two million children have been using various forms of assisted human reproduction, such as in vitro fertilization, donor eggs, donor sperm, and surrogate mothers.[13]

Other reproductive technologies include intracytoplasmic sperm injection—that is actually inserting a sperm into an egg to cause fertilization. This technique is used in cases where the father's sperm lacks the ability to breach the egg's protective outer coating without help. Thousands of children have been born using this technique.

Another advance is oocyte cytoplasm transfer in which cytoplasm from donor eggs is injected into a patient's eggs that have had difficulty developing, possibly because of defects in their energy-producing cellular organelles called mitochondria. This is actually a kind of germline intervention since the mitochondrial genes are transferred from the donor. So, in a sense, a child born using this technique has three genetic parents. Several dozen healthy children have been born using this technique so far.

Then there is pre-implantation genetic diagnosis—in which parents create embryos that are then screened for genetic diseases and only those found free of the diseases are implanted in the mother's womb. In March 2002, the *Journal of the American Medical Association* published articles highlighting the case of an anonymous married thirty-year-old geneticist who will almost certainly lose her mind to early-onset Alzheimer's disease by age 40. She chose to have her embryos tested in vitro for the disease gene and then implanted only embryos without the gene into her womb. The result was the birth of a healthy baby girl about a year ago—one who will not suffer Alzheimer's in her forties. The mother in this case certainly knows what would face any child of hers born with the disease gene. Her father, a sister, and a brother have all already succumbed to early Alzheimer's.

To achieve this miracle, the mother used the services of the Reproductive Genetics Institute (RGI) in Chicago, a private fertility clinic that has pioneered this kind of testing, called pre-implantation genetic diagnosis (PGD).

PGD is being used by more and more parents who want to avoid passing on devastating genetic diseases to their progeny. Diseases tested for include cystic fibrosis, Tay-Sachs, various familial cancers, early-onset Alzheimer's, sickle cell disease, hemophilia, neurofibromatosis, muscular dystrophy, and Fanconi anemia.

The prospect of using PGD for any diagnosable disease is particularly upsetting to some critics. "Today it's early-onset Alzheimer's. Tomorrow it could easily be intelligence, or a good piano player or many other things we might be able to identify the genetic factors for," said Jeffrey Kahn, director of the University of Minnesota's Center for Bioethics. "The question is, whether we ought to."[14] In the *Washington Post*, Kahn insisted, "It's a social decision. This really speaks to the need for a larger policy discussion, and regulation or some kind of oversight of assisted reproduction."[15]

But there is no reasonable ethical objection for using genetic testing to avoid disease. That's what medicine is supposed to do—cure and prevent disease. Kahn is right that parents will someday use PGD to screen embryos for desirable traits such as tougher immune systems, stronger bodies, and smarter brains. It is hard to see what is ethically wrong with parents taking advantage of such testing, since it is aimed at conferring general benefits that any child would want to have.

The usually sensible University of Pennsylvania bioethicist Arthur Caplan asked, "Testing for diseases that are going to appear 30 or 40 years from now, does that make any sense, since people are mortal?"[16] One might think that a mother whose family has been grievously afflicted with this disease and who faces it herself is in a better position to decide than even the most brilliant academic bioethicist is. Surely it is the height of moral rectitude for a parent to spare her children the terrible fate that this mother knows lies in store for her. If it's alright to use efficacious medical treatments to cure a forty-year-old with Alzheimer's, it's alright to prevent him from getting it in the first place.

Caplan added that PGD procedures are certain to prove highly controversial "because that's really getting into designing our descendants."[17] But what horrors do such designer babies face? Longer, healthier, smarter, and perhaps even happier lives? It is hard to see any ethical problem with that. The good news is that bioconservatives have so far failed to outlaw PGD and more than 2,500 disease-free children have been born using PGD.

Many bioconservatives favor establishing a government agency with the power to impose ethical rulings on people's use of new biotechnologies. Their model is the United Kingdom's Human Fertilisation and Embryology Authority (HFEA). Heretofore, government has asserted regulatory authority on the grounds of insuring the safety, quality, and efficacy of new products and techniques. The HFEA rules not only on the safety and efficacy of new techniques, but also on their moral acceptability.

Consider the case of the Whitaker family from Sheffield, England, to see just how dangerous it is to allow a government agency to interfere in a family's reproductive decisions. In 2002, Michelle and Jayson Whitaker asked the HFEA for permission to use in vitro fertilization and PGD to produce a tissue-matched sibling for their son Charlie, who suffers from a rare anemia. That disease caused him to need a blood transfusion every three weeks. The HFEA refused, calling the procedure "unlawful and unethical," ruling that tissue matching is not a sufficient reason to attempt embryo selection.

Desperate, the Whitakers came to the United States, where PGD is still legal. In June 2003, Michelle Whitaker gave birth to James, whose umbilical cord stem cells are immunologically compatible with Charlie's. The stem cells have now been transplanted. His doctors report that Charlie's bone marrow looks normal and he appears to be cured.[18] Please keep in mind that taking stem cells from James's umbilicus in no way endangered or harmed him.

HFEA's rejection of the Whitakers' request was not based on safety or efficacy, but on the moral opinions of the authority's governing panel. Such a regulatory authority necessarily turns differences over morality into win/lose propositions, with minority views—and rights—overridden by the majority.

Besides the biomedical procedures discussed here, the future may hold artificial chromosomes that contain traits selected by parents and inserted in the fertilized egg before it begins dividing. Such artificial chromosomes may well have genetic docking stations so that new genes can be inserted later in adulthood.

Biotechnological progress is not confined to reproduction. One of the more controversial areas of research is what has been called therapeutic cloning or research cloning—aimed eventually at the creation of embryos using a patient's own DNA for the purpose of deriving embryonic stem cells. It is thought that such stem cells would be perfect transplants for patients. The National Academy of Sciences estimates that one hundred million Americans could one day be candidates for embryonic and adult stem cell treatments.[19] The House of Representatives has already twice passed a bill that would criminalize this research and any American who goes abroad to take advantage of such therapies—the penalty is ten years in prison and $1 million in fines. Kansas Republican senator Brownback is proposing a similar bill in the Senate. The majority of the President's Council on Bioethics voted in favor of a moratorium on research on therapeutic cloning to produce stem cells and voted unanimously to ban reproductive cloning forever.

Those who are afraid of posthumanity regularly invoke the dystopian visions of Mary Shelley's *Frankenstein*, Aldous Huxley's *Brave New World*, and C. S. Lewis's *The Abolition of Man* as warnings of what a future of unleashed human biotechnology might hold. Fortunately, they are wrong.

Why? First, many opponents of human genetic engineering are either conscious or unconscious genetic determinists. They fear that biotechnological knowledge and practice will somehow undermine human freedom. In a sense, these genetic determinists on the Left and the Right believe that somehow human freedom resides in the gaps of our knowledge of our genetic makeup. If parents are allowed to choose their children's genes, according to left-wing environmentalist Bill McKibben in his anti-biotech book, *Enough: Staying Human in an Engineered Age*, then they will have damaged their children's autonomy and freedom. In other words, for these bio-fundamentalists "Freedom is Ignorance." They apparently believe that our freedom and autonomy depend on the unknown and random combinations of genes that a person inherits. But even if they were right—and they are not—genetic ignorance of this type will not last. Human genome sequencer Craig Venter predicts that whole genome testing will become available by 2010 so that every person's entire complement of genes can be scanned and known at his or her physician's office. Once whole genome testing is perfected we will all learn even what our randomly conferred genes may predispose us to do. Human freedom will then properly be seen as acting to overcome these predispositions, much like a former alcoholic can overcome his thirst for booze. Fortunately biotechnology will help here as well with the development of neuropharmaceuticals to enhance our cognitive abilities and change our moods.

Second, biotech opponents cite C. S. Lewis's argument that one decisive generation that first masters genetic technologies will control the fate of all future generations. But when has it not been true that past generations control the genetic fate of future generations? Our ancestors, too—through their mating and breeding choices—determined for us the complement of genes that we all bear today. They just didn't know which specific genes they were picking. Fortunately, our descendants will have at their disposal ever more powerful technologies and the benefit of our own experiences to guide them in their future reproductive decisions. In no sense are they prisoners of our decisions now. *Of course, there is one case in which future generations would be prisoners of our decisions now, and that's if we fearfully elect to deny them access to the benefits of biotechnology and safe genetic engineering*. Biotech opponents like Kass and Callahan are wrong—the future will not be populated by robots who may look human but who are unable to choose for themselves their own destinies—genetic or otherwise.

Bioconservatives oppose the use of "enhancement technologies to shape the destiny of others, and especially their children."[20] Such parents, we are told by conservative writer Dinesh D'Souza, are "totalitarians" engaging in "despotism" and "tyranny." Those of us who see no moral objection to genetic enhancements, we are told again by D'Souza, "speak about freedom and choice, although what [they] advocate is despotism and human bondage."[21] Left-winger Bill McKibben agrees writing, "The person left without any choice *at all* is the one you've engineered."[22] This is complete nonsense.

Again, this is an expression of hard genetic determinism that is simply not warranted by biology. A gene that enhances one's capacity for music doesn't mean that its possessor must become another Scott Joplin or Keith Jarrett; genes simply don't work that way. All of us have many capacities stemming from his or her specific genetic endowment. Perhaps, I could have become a professional basketball player or a computer engineer, but I chose not to develop those particular abilities despite the fact that my specific complement of genes might have allowed me to do so.

It is certain that it is our individual brains, and not our genes, that make us individual human beings. The case of identical twins proves the point: They have exactly the same genes, but they are different, sometimes very different, people. That's why, in recent years, our society has legally defined death as brain death. Once our brains are gone, we are gone, even though our bodies—with all their genes—may live on. The fact is, we respect people, not their genes. In a very real sense, we are no longer at the mercy of genes; instead our genes are at the mercy of our brains.

Giving children such enhanced capacities as good health, stronger bodies, and more clever brains, far from constraining them, would in fact give them greater freedom and more choices. It's surely a strange kind of despotism that enlarges a person's abilities and options in life.

Another often-heard objection is that genetic engineering will be imposed on "children-to-be" without their consent. First, do I need to remind anyone that *not one of us* gave our consent to be born, much less to be born with the specific complement of genes that we bear. Thus the children born by means of assisted reproductive therapies and those produced more conventionally stand in exactly the same ethical relationship to their parents.

Secondly, the absurdity of the bioconservative requirement for prenatal consent becomes transparent when you ask bioconservatives if they would forbid fetal surgery to correct spina bifida or fetal heart defects. After all, those fetuses can't give their consent to those procedures, yet it is certainly the moral thing to do. For that matter, taking the bioconservative position on consent to its logically absurd conclusion would mean that children couldn't be treated with drugs, or receive vaccinations. So using future biotechnical means to correct genetic diseases like cystic fibrosis or sickle cell anemia at the embryonic stage will similarly be a morally laudatory activity. Surely one can assume that the beneficiary—the not-yet-born, possibly even the not-yet-conceived child—would happily have chosen to have those diseases prevented.

Now let's say a parent could choose genes that would guarantee her child a 20-point IQ boost. It is reasonable to presume that the child would be happy to consent to this enhancement of her capacities. How about plugging in genes that would boost her immune system and guarantee that she would never get colon cancer, Alzheimer's, AIDS, or the common cold? Again, it seems reasonable to assume consent. These enhancements are general capacities that any human being would reasonably want to have. In fact, lots of children already do have these capacities naturally, so it's hard to see that there is any moral justification for outlawing access to them for others.

In a joint op-ed in the *Los Angeles Times*, left-winger and longtime biotechnology opponent Jeremy Rifkin and right-wing editor of the *Weekly Standard*, William Kristol declare, "Humans have always thought of the birth of their children as a gift bestowed by God or a beneficent nature."[23] Unlike Rifkin and Kristol, I do not claim to know the mind of the Creator. However, I do wonder why they believe that the Creator doesn't want us to use the intelligence he gave humanity to discover biomedical interventions to relieve suffering, cure diseases, and improve lives? Instead of submitting to the tyranny of Nature's lottery—which cruelly deals out futures blighted with ill health, stunted mental abilities, and early death—one could certainly argue that God wants parents to be able to open more possibilities for their children to have fulfilling lives. Genetic enhancements to prevent these ills would not violate a child's liberty or autonomy and certainly do not constitute a form of genetic slavery as some opponents claim. In fact, through the gift of technology brought about by human intelligence, parents can increasingly bestow not only the gift of life, but also the gift of good health, on their children.

Kristol and Rifkin darkly warned of the advent of "a commercial eugenics civilization," offering "a new form of reproductive commerce with frightening implications for the future of society."[24] They are fabricating a bogus ethical dichotomy pitting "utilitarians" against those who allegedly "believe in the intrinsic value of human life." Despite their invidious moral posturing, Kristol and Rifkin do not have a lock on ethical rectitude. The intrinsic value of human life is a given for all sides in this debate. The battle is really between those who want to use the gifts of human reason and human compassion to ameliorate illness and death and those like Kristol and Rifkin who counsel fatalistic acceptance of the manifold cruelties randomly meted out by nature.

Let's turn again to Dinesh D'Souza, who claims, "The power they seek to exercise is not over 'nature,' but over other human beings."[25] Actually, most of those who want access to genetic technologies for their children are motivated by exactly the opposite desire: What they seek is the power to defend their children against the manifold cruelties and indignities that "nature" so liberally dispenses, and thus make it possible for their children to have fulfilling lives. The good news is that any would-be tyrannical parents who buy into the bioconservatives' erroneous notions of hard genetic determinism will be disappointed. Their children will have minds and inclinations all distinctly their own, albeit genetically enhanced.

Others like Bill McKibben and Adam Wolfson, the former editor of *The Public Interest*, raise egalitarian worries that the availability of enhancement technologies will create two classes in society. "The political equality enshrined in the Declaration of Independence can't withstand the destruction of the idea that humans are in fact equal," claims Bill McKibben.[26] These opponents of biomedical progress make the mistake that the ideal of political equality rests on the notion of actual physical and mental equality. That's nonsense. The ideal of political equality arose from the Enlightenment's insistence that since no one has access to absolute truth, no one has a moral right to impose his or her values and beliefs on others. Political equality has never rested on claims about human biology. We all had the same human biology during the long millennia in which slavery, patriarchy, purdah, and aristocratic rule were social norms.

Besides, their view is shortsighted. The type of genetic interventions contemplated here will likely become available to more and more people as wealth increases in our society and as their relative prices go down. So ever cheaper genetic medical technologies seem to me to be a recipe for *eliminating genetic inequalities rather than perpetuating them*. In any case, the ideals of democracy and political equality are sustained chiefly by the principle that people are responsible moral agents who can distinguish right from wrong and therefore deserve equal consideration before the law and a respected place in our political community. The broad ability to distinguish right from wrong does not depend on the genetics of IQ, skin color, or gender. With respect to political equality, genetic differences are differences that make no difference. Having some citizens who take advantage of genetic technologies and others who do not does not alter that principle.

But something worse than mere genetic engineering fills McKibben "with blackest foreboding." What is it? The prospect of physical immortality, of course. "It would represent, finally, the ultimate and irrevocable divorce between ourselves and everything else," he asserts. "The divorce, first of all, between us and the rest of creation."[27] Left-wing bioconservative McKibben is not alone in his "fear of life." The idea that biotechnological progress might dramatically increase healthy human life spans horrifies this whole strange biopolitical alliance.

Leon Kass has asserted, "The finitude of human life is a blessing for every individual, whether he knows it or not."[28] Daniel Callahan, founder of the Hastings Center, has declared, "There is no known social good coming from the conquest of death." Callahan added, "The worst possible way to resolve [the question of life extension] is to leave it up to individual choice."[29] When asked if the government has a right to tell its citizens that they have to die, Fukuyama answered, "Yes, absolutely."[30]

Again, it will not be as easy to retard aging as we might hope, but even the President's Council on Bioethics acknowledges, in its report *Beyond Therapy: Biotechnology and the Pursuit of Happiness*, that "It seems increasingly likely that something like age-retardation is in fact possible."[31]

Why do we want to stay married to Nature anyway? She has certainly been an inconstant wife, liberally afflicting us with nasty surprises like birth defects, diseases, earthquakes, hurricanes, famines, and so forth. Actually an amiable separation might be good for both Nature and humanity. The less we depend on Nature for our sustenance, the less harm we do her. Setting that aside, why does McKibben believe that death is good for us? "Without mortality, no *time*. All moments would be equal; the deep, sad, human wisdom of Ecclesiastes would vanish. If for everything there is an endless season, then there is also no right season," writes McKibben. "The future stretches before you endlessly flat."[32]

Actually, the deep, sad wisdom of Ecclesiastes is a very powerful human response of existential dread to the oblivion that stretches endlessly before the dead. "For the living know that they shall die: but the dead know not any thing, neither have they any more a reward; for the memory of them is forgotten. Also their love, and their hatred, and their envy, is now perished; neither have they any more a portion forever in any thing that is done under the sun," writes the Preacher.[33] I would like to suggest that if death is not inevitable, most of humanity will be happy to spend the extra time granted them to learn new teachings and new wisdom.

If, however, the endless future turns out to be as horrible as McKibben imagines it to be, then people will undoubtedly choose to give up their empty meaningless lives. On the other hand, if people opt to live yet longer, wouldn't that mean that they had found ample enough pleasure, joy, love, and yes, meaning to keep them going? McKibben's right, we don't know what immortality would be like—but should that happy choice become available through biomedical progress, each of us can still decide whether we want to enjoy it or not. Besides even if the ultimate goal of the biotechnological quest is immortality, what will be offered immediately will only be longevity. The experience of longer lives will give humanity an opportunity to see how it works out. If longer lives do become meaningless, then surely people will not choose longer meaningless lives. In a sense, if immortality is a problem, it is a self-correcting problem. Death always remains an option.

Each of these pro-death policy intellectuals has their various reasons to favor human mortality, but bioethicist Daniel Callahan's views are fairly representative. In the June 2004 issue of the *Journal of Gerontology*, bioethicist Daniel Callahan makes three arguments.

First, he points out that the "problems of war, poverty, environment, job creation, and social and familial violence" would not "be solved by everyone living a much longer life." Second, he asserts that longer lives will lead mostly to more golf games not new social energy. "I don't believe that if you give most people longer lives, even in better health, they are going to find new opportunities and new initiatives," Callahan writes. And thirdly, Callahan is worried about what longer lives would do to child bearing and rearing, social security and Medicare. Finally he demands, "Each one of the problems I mentioned has to be solved in advance. The dumbest thing for us to do would be to wander into this new world and say, 'We'll deal with the problems as they come along.'"[34]

Callahan's first argument is a non sequitur. People already engage in lots of activities that do not aim directly at "solving" war, poverty, environment, job creation, and the rest. Surely we can't stop everything until we've ended war, poverty, and familial violence. Anti-aging biomedical research wouldn't obviously exacerbate any of the problems listed by Callahan and might actually moderate some of them. If people knew that they were likely to enjoy many more healthy years they might be more inclined to longer-term thinking aimed at remedying some of the problems mentioned by Callahan. Besides, as Case Western Reserve bioethicist Stephen Post has correctly observed: "If we were to insist that technological developments of all sorts wait until the world becomes perfectly just, there would be absolutely no scientific progress."[35]

Second, Callahan's "longer life equals more golf" argument is not only condescending, it ignores the ravages that physical decline visits on people. Surely Callahan, at age 74, sees a lack of "new energy" among his confreres. Even if people are healthy at age 75, their "energy" levels will be lower than they were when they were thirty. They may not begin "new initiatives" because they can't expect to live to see them come to fruition. But diminishing physical energy isn't the only problem; there is also waning psychic energy. "There's a factor that has nothing to do with physical energy. That is the boredom and repetition of life," he argues. He added, "I ran an organization for 27 years. I didn't get physically tired. I just got bored doing the same thing repetitiously."[36]

My first question is, "Dan, why didn't you change jobs?" It doesn't seem reasonable to conclude that just because Callahan is bored with life, that we all will become so. Modern material and intellectual abundance is offering a way out of the lives of quiet desperation suffered by many of our impoverished ancestors. The twenty-first century will offer an ever-increasing menu of possible life plans and choices. Surely exhausting the coming possibilities will take more than one standard lifetime to achieve. Besides even if you do want to play endless games of golf and can afford it, why is that immoral? And if you become bored with life and golf, no one is forcing you to hang around.

Doubling healthy human life expectancy would create some novel social problems, but would they really be so hard to deal with? Callahan cites the hoary example of brain- dead old professors blocking the progress of vibrant young researchers by grimly holding onto tenure. That seems more of a problem for medieval holdovers like universities than for modern social institutions like corporations. Assuming that it turns out that even with healthy long-lived oldsters that there is some advantage for turnover in top management, then corporations that adopt that model will thrive and those that do not will be out competed. Besides, even today youngsters don't simply wait around for their elders to die—they go out and found their own companies and other institutions. Bill Gates didn't wait to take over IBM; he founded Microsoft at age 20. Nor did human genome sequencer Craig Venter loiter about until the top slot at the National Institutes of Health opened up. Sergey Brin and Larry Page founded Google, currently the world's most popular internet search engine company, in their twenties. Michael Dell founded Dell Computers, the world's largest seller of personal computers, at age 19.

In politics, besides the presidency, we already limit the terms of many state and local offices. And empirical evidence cuts against Callahan's worries that healthy geezers will slow economic and social progress. After all, social and technological innovation has in fact been most rapid in those societies with the highest average life expectancies. Yale economist William Nordhaus argues that the huge increase in average life expectancy since 1900 from forty-seven years to seventy-seven years today has been responsible for about half the increase in our standard of living in the United States.[37]

Even more weirdly, Callahan is worried about Social Security and Medicare. His failure of imagination is breathtaking in this regard—folks will be chronologically older, but not physically frail. Thus they will be expected to continue to be productive and support themselves. Assuming that age-retardation is possible, the many illnesses and debilities that accompany aging will be postponed. If one is going to live to be 140, one has a lot of time to plan and save for the future.

But what about child bearing and child rearing? The assumption here seems to be that healthy long-lived oldsters would be less interested in reproducing. A first response might be, so what? Shouldn't the decision to have children be up to individuals? After all, it is already the case that the countries with the highest life expectancies are the ones that have the lowest levels of fertility. This lack of interest in progeny would have the happy side effect of addressing some concerns that doubling human life spans might lead to over-population. No one can know for sure, but it could well be that bearing and rearing children would eventually interest long-lived oldsters who would come to feel that they had the time and the resources to do it right.

Callahan's final demand that all the problems that might be caused by doubled healthy life spans be solved in advance is just silly. Humanity did not solve all of the problems caused by the introduction of farming, electricity, automobiles, antibiotics, sanitation, and computers in advance—we proceeded by trial and error and corrected problems as they arose. We should be allowed do the same thing with any new age-retardation techniques that biomedical research may develop and we'll be happy to do so.

Meanwhile, Leon Kass has argued, "immortality is a kind of oblivion—like death itself."[38] He adds:

> Mortality as such is not our defect, nor is bodily immortality our goal. Rather mortality is at most a pointer, a derivative manifestation, or an

> accompaniment of some deeper deficiency . . . the human soul yearns for, longs for, aspires to some condition, some state, some goal toward which our earthly activities are directed but which cannot be attained in earthly life. . . . Man longs not so much for deathlessness as for wholeness, wisdom, goodness, and godliness.[39]

Taking Kass on his own terms, what can he mean by asserting that immortality would be a kind of oblivion? Keep in mind: Whatever the risks of "oblivion" posed by immortality, they must surely be balanced against the certain oblivion of the grave.

Kass points to the frivolity of the lives of the gods in Greek myths. Nothing is serious to them because eternity stretches endlessly before them. Their choices are whimsical and capricious. They are not strictly speaking moral beings because no decisions or choices are final—among themselves the passage of time can reverse all harms and all insults. It's all just a game and mortals are of no more real significance to them than are mosquitoes to us.[40]

Do these myths provide a warning to heedless humanity about the pursuit of immortality? Perhaps not. After all, even an immortal being can decide to become mortal. For example, in Christianity, God himself became a man and He died. To the extent that the oblivion of death gives urgency and poignancy to life—it will still do so even for very long-lived people. It is noble now to give one's life to save others, but how much more noble it will be when defenders sacrifice centuries rather than decades of their lives for others.

Another problem with the Greek gods as cautionary models is that they existed in an eternity with no progress, no goals other than playing the same endless round of games among themselves. Many visionaries who argue for dramatically extending human life spans foresee a day when people can use technology to become very much like gods, with vast new abilities to apprehend new knowledge and to achieve new goals. But for a very long time, such godlike people will clearly be neither omniscient nor omnipotent—so there will remain plenty of adventure and wisdom to seek in the wider universe.

What about Kass's claim that "Man longs not so much for deathlessness as for wholeness, wisdom, goodness, and godliness"? This is again simply a false dichotomy. One can long for deathlessness *and* wholeness, *and* wisdom, *and* goodness, *and* godliness. Or is Kass saying that we seek after these virtues only because we are mortal? If so, it cannot be that they are somehow intrinsically or transcendently important. In any case, longer lives allow us to seek further for those human virtues Kass says he prizes. Defeating death is not the final act in our technological or spiritual quest—it is merely a vital prelude to our further pursuit of wholeness, wisdom, goodness, and yes, even godliness. Surely, for all too many people, perhaps for most of us, the pursuit of these virtues has been or will be cut short by death well before we have managed to attain them.

Those anxious about the advance of biotechnology offer up a wide variety of other horror stories that they hope will scare citizens and legislators into criminalizing certain biotech research. Francis Fukuyama warns that humanity's Nietzschean will-to-power would tempt us to create subhuman slaves. Specifically, Fukuyama suggested that biotechnology might be used to create half human-half chimpanzee slaves with the intelligence of a twelve-year-old boy.[41]

I would like to suggest that Fukuyama is overlooking a few practical concerns, like the fact that mothers willing to bear "subhuman slaves" in their wombs are likely to be scarce. And who would want a subhuman slave anyway? Fully human slaves don't appear to work out so well in the modern world. If you want real travel efficiency, you don't call for a slave-carried palanquin. You get into your Dodge Neon. If you need to write a letter, you don't summon your scribe. You fire up your Dell laptop.

Who doubts that ever-more efficient and obedient machines will be cheaper and more practical solutions to the "servant problem" than any half human-half chimp slave would be? Besides anyone who has ever tried to supervise the activities of a twelve-year-old for any length of time would likely pass on the opportunity to own one of Fukuyama's subhuman slaves.

The question is, does biotechnology pose any novel moral concerns? The general answer is, no. It is wrong to diminish the health or mental abilities of a child today—we call that child abuse and we outlaw it. We don't allow slavery, even for twelve-year-olds. Biomedical progress will not change these bedrock moral principles.

Do we need regulations, vast new federal agencies aimed at controlling genetic technologies? Again, no. Instead of setting up an intrusive and inevitably sluggish regulatory system in an effort to prevent abuses in advance, as Fukuyama wants, the better way to proceed is to allow new technologies to develop, and then ameliorate problems as they arise. Rather than create a barrier that timidly allows only a few approved technologies through, we should instead encourage technological progress and correct problems as they become evident.

This is the way that physicians have developed and deployed the highly successful and innovative fertility treatments that have helped hundreds of thousands of couples to have families. When a rare abuse is uncovered, law enforcement has had no difficulty in punishing malefactors in the fertility industry. For example, take the case of fertility doctors in California who swapped embryos without their patients' consent. And also the case of a fertility specialist in Virginia who artificially inseminated his patients with his own sperm. Some have argued that such cases prove we need federal regulations, but actually they prove just the opposite. In the California case, the doctors were indicted and convicted of fraud.[42] In the Virginia case, the fertility specialist was convicted on fifty-two counts of fraud and perjury and sent to prison for five years.[43]

Clearly our legal system can already punish such criminals without new federal rules or regulatory bodies. Similarly, most problems arising from the further development of genetic technologies can be dealt with by applying already existing laws and regulations like those prohibiting fraud and child abuse.

It is true that there are some pernicious ideas lurking in some quarters of the intellectual Left, for example, mandatory government-subsidized eugenics in the name of equality. Leftist thinker Ronald Dworkin is a supporter of such a project.[44] This elitist egalitarian impulse, not biotechnology, is the real threat. But instead of condemning a pernicious idea, some biotech opponents' fears of how egalitarians could misuse biotechnology drive them illogically to condemn the technology as well. That is somewhat akin to arguing that simply because airplanes can be rammed into buildings, we should therefore ban jetliners.

It is true for genetic engineering, as for all other technologies, that some people will misuse it; tragedies will occur. Given the sorry history of government-sponsored eugenics, it is vital that control over genetic engineering must never be given to any governmental authority. But to use biotechnology and genetic engineering is not, by definition, to abuse it. This technology offers the prospect of ever-greater freedom for people, and should be welcomed by everyone who cares about human happiness and human flourishing.

People who take advantage of the fruits of biotechnological progress in the future will be neither Frankenstein monsters nor genetic robots. Rather, they will be our grateful descendants for whom we have eased the burdens of disease, disability, and early death, if only we choose not to slow or kill the development of this new technology. They will look back in wonder, and perhaps in horror, at those who would have denied them the blessings of biomedical progress.

In his famous essay "The End of History?" Fukuyama declared that our generation is witnessing "the end point of mankind's ideological evolution and the universalization of Western liberal democracy as the final form of human government." Fair enough. But for Fukuyama, the end of history is a "sad time" because "daring, courage, imagination, and idealism will be replaced by economic calculation." Also, he claims, "in the post-historical period there will be neither art nor philosophy, just the perpetual caretaking of the museum of human history."[45] How ironic that Fukuyama now spends his time demonizing biotechnological progress and transhumanism, a nascent philosophical and political movement that epitomizes the most daring, courageous, imaginative, and idealistic aspirations of humanity.

In a sense, the battle over the future of biotechnology—and, if Fukuyama is correct, the future of humanity—is between those who fear what humans, having eaten of the Tree of Knowledge of Good and Evil, might do with biotechnology and those of us who think that it is high time that we also eat of the Tree of Life.

Notes

1. Jeremy Rifkin, "Fusion BioPolitics," *The Nation*, February 18, 2002, http://www.thenation.com/doc/20020218/rifkin.

2. Jeremy Rifkin, "The Price of Life," *Guardian Unlimited*, November 15, 2000, http://www.guardian.co.uk/comment/story/0,3604,397765,00.html.

3. Jeremy Rikfin, cited by Ronald Bailey, "Rage Against the Machines," *Reason*, July 2001, http://www.reason.com/0107/fe.rb.rage.shtml.

4. Bill McKibben, *Enough: Staying Human in an Engineered Age* (New York: Henry Holt and Company, 2003), 186.

5. George Annas et al., "Protecting the Endangered Human: Toward an International Treaty Prohibiting Cloning and Inheritable Alterations," *American Journal of Law and Medicine*. 28:2, 3 (2002): 151–178.

6. Tom Athanasiou and Marcy Darnovsky, "The Genome as Commons," *World-Watch*, July/August 2002, http://www.genetics-and-society.org/resources/cgs/200207_worldwatch_darnovsky.html.

7. Richard Hayes, "The Science and Politics of Genetically Modified Humans," *WorldWatch*, July/August 2002, http://www.genetics-and-society.org/resources/cgs/200207_worldwatch_hayes.html.

8. Daniel Callahan, *Setting Limits: Medical Goals in an Aging Society* (New York: Simon and Schuster, 1987).

9. Leon R. Kass, "The Moral Meaning of Genetic Technology," *Commentary* 108 (September 1999): 33.

10. Leon R. Kass, "The Wisdom of Repugnance," *New Republic*, June 2, 1997, http://www.catholiceducation.org/articles/medical_ethics/me0006.html.

11. Francis Fukuyama, "In Defense of Nature, Human and Non–Human," *New Perspectives Quarterly*, July 16, 2002, http://www.digitalnpq.org/global_services/global%20viewpoint/07-16-02.html.

12. Adam Wolfson, "Politics in a Brave New World," *The Public Interest*, Winter 2001, http://findarticles.com/p/articles/mi_m0377/is_2001_Wntr/ai_69411628/pg_7.

13. Personal communication, September 2004.

14. Jeffrey Kahn cited in "Not So Slippery Slope," *Washington Post*, March 3, 2002: B06.

15. Jeffrey Kahn cited by Rick Weiss, "Alzheimer's Gene Screened from Newborn," *Washington Post*, February 27, 2002: A01.

16. Arthur Caplan cited by the Associated Press, "Marked Mom Births Clean Baby," February 26, 2002, http://www.wired.com/news/medtech/1,50693-0.html.

17. Ibid.

18. John Crowley, "'Designer Baby' Cures Brother in Stem Cell Breakthrough," *Daily Telegraph*, October 21, 2004: 3; Antonia Hoyle, "My Little Brother Was Born to

Save My Life; Exclusive: Charlie, 6, Gets Health All-Clear in Medical 'First,'" *The Mirror*, August 16, 2005: 12–13.

19. National Academy of Sciences, *Stem Cells and the Future of Regenerative Medicine* (Washington, DC: National Academy Press, 2002).

20. Dinesh D'Souza, "Staying Human: The Danger of Techno-Utopia," *National Review*, January 22, 2001, http://www.findarticles.com/p/articles/mi_m1282/is_1_53/ai_69240190.

21. Ibid.

22. McKibben, *Enough*, 191.

23. William Kristol and Jeremy Rifkin, "First Test of the Biotech Age: Human Cloning," *Los Angeles Times*, March 6, 2002: Part 2, 11.

24. Ibid.

25. D'Souza, "Staying Human."

26. McKibben, *Enough*, 40.

27. Ibid., 160.

28. Leon R. Kass, "*L'Chaim* and Its Limits: Why Not Immortality?" *First Things*, May 2001, http://www.firstthings.com/ftissues/ft0105/articles/kass.html.

29. Daniel Callahan cited by Ronald Bailey, "Intimations of Immortality," *Reason*, March 6, 1999, http://reason.com/opeds/rb030600.shtml.

30. Francis Fukuyama cited by George Dvorsky, "Deathist Nation," *Betterhumans.com*, June 6, 2004, http://archives.betterhumans.com/Columns/Column/tabid/79/Column/266/Default.aspx.

31. See President's Council on Bioethics, *Beyond Therapy: Biotechnology and the Pursuit of Happiness*, October 2003, http://www.bioethics.gov/reports/beyondtherapy/chapter4.html.

32. McKibben, *Enough*, 159.

33. Ecclesiastes 9:5–6.

34. See Gregory Stock and Daniel Callahan, "Debates: Point-Counterpoint: Would Doubling Human Life Span Be a Net Positive or a Negative for Us Either as Individuals or as a Society?" *Journal of Gerontology: Biological Sciences* 59 (2004): http://biomed.gerontologyjournals.org/cgi/content/full/59/6/B554.

35. Stephen Post, "Debates: Point-Counterpoint: Would Doubling Human Life Span Be a Net Positive or a Negative for Us Either as Individuals or as a Society?" *Journal of Gerontology: Biological Sciences* 59 (2004): http://biomed.gerontologyjournals.org/cgi/content/abstract/59/6/B534.

36. Stock and Callahan, "Debates: Point-Counterpoint," 2004.

37. William Nordhaus, "The Health of Nations: The Contribution of Improved Health to Living Standards," in *The Economic Value of Medical Research*, ed. Kevin Murphy and Robert Topel (Chicago: University of Chicago Press, 2002).

38. Leon R. Kass, "The Case for Mortality," *The American Scholar* 52 (Winter 1977–1978): 185.

39. Ibid., 185, 187.

40. Ibid., 184–185.

41. See Francis Fukuyama cited by Ronald Bailey, "Right-Wing Biological Dread," *Reason*, December 12, 2001, http://www.reason.com/rb/rb121201.shtml.

42. See "Fertility Scandal Doctor Sentenced," Associated Press, May 12, 1998.

43. See Bill Miller, "Fertility Doctor Begins Prison Term: Va. Man Who Fathered Patients' Children Nears End of Appeals," *Washington Post*, February 19, 1994: B3.

44. Ronald Dworkin, *Sovereign Virtue: The Theory and Practice of Equality* (Cambridge, Mass.: Harvard University Press, 2000), 427–452.

45. See Francis Fukuyama, "The End of History?," *National Interest* 16 (Summer 1989): 18.

THREE

Bioethics and Human Betterment

Have We Lost Our Ability to Dream?

RONALD M. GREEN

I want to begin with some statistics. If you were born in 1900, it is likely that you would never have survived infancy. For every one thousand births, a hundred infants died before their first birthday. In some U.S. cities in 1900, up to 30 percent of all newborns died in the first year of life. Today, the U.S. infant mortality rate is 7.2 per 1,000, more than a 90 percent decline since 1900.

Maternal mortality was also very high in 1900, averaging 6–9 maternal deaths per 1,000 live births—nearly a 1 percent chance of dying in childbirth. The corresponding figure today is 7.7 deaths per 100,000 live births, a 99 percent decline in the course of a century.[1]

Taking this very high infant mortality rate into account, average life expectancy at birth for white females in 1900 was fifty years. If a girl managed to reach age 10, she could count on an additional fifty-two years of life, dying on average at age 62.[2]

Today, average life expectancy for a white female at birth is eighty years. At age 10, she can count on an additional seventy-one years of life. This represents a 31 percent increase in longevity since 1900.[3] The averages are somewhat lower but also greatly improved for white males and non-white males and females.[4]

Beyond these quantitative measures of health improvement, there are many qualitative achievements worth celebrating. When a person in midlife experiences blockage in the cardiac arteries, a leading cause of death just thirty years ago, new interventions are available that can restore function and provide additional years or even decades of active life. Hip joint replacements offer mobility to people who previously were consigned to wheelchairs or walkers. Assisted Reproductive Technologies (ARTs) have allowed hundreds of thousands of infertile couples to have children of their own. Cosmetic medicine also contributes to human well-being. New techniques in plastic surgery help individuals to overcome blemishes caused by accident or birth defects. Prozac and related psychotropic medications give life back to many who were crippled by depression or anxiety.

These achievements are the work of people in the fields of medicine, public health, and basic science who were not afraid to dream. Their dreams were of human life progressively freed from ancient scourges of disease and disability. Some imagined that human beings could even be improved: that we could live longer than the historical norm, that we could become more resistant to disease or disability, and that we might even develop new powers of body and mind.

Sadly, we have entered a period in history marked by a loss of nerve. Biomedical dreaming has gone out of fashion. In its place, voices of caution have come to predominate in public discourse. From leading bioethics think tanks to a host of popular books and articles, the message is "Enough!" Appeals for moratoriums on research are the order of the day, often fed by separate and independent agendas. They come from some who, on religious grounds, oppose human embryo or stem cell research. Others, hearkening back to Nazi abuses or opposed to the use of genetically modified organisms in agriculture, object to advances in genetics, reproductive medicine, and prenatal testing. In portions of Europe, especially Germany, these sentiments sometimes reach the level of what one writer calls an "anti-technological hysteria."[5]

In Europe and the United States, a host of young people enjoying a greater health status than any generation in history, but often unaware of how these gains were accomplished, tacitly support the oppositional voices. Justifiable worries about environmental degradation and nuclear proliferation foster a generalized attitude of distrust of science and technology. Together, all these forces are producing an informal but powerful new coalition of opposition that has already begun to impede biomedical research in some political jurisdictions.

Late in 2003, the President's Council on Bioethics published a report entitled *Beyond Therapy: Biotechnology and the Pursuit of Happiness*.[6] The report examines various biomedical technologies in the areas of genetics, life extension, and neurological pharmacology that hold out the promise not just of curing disease but of improving human biology to a state that is "better than well." The chapter dealing with life extension technologies is disturbingly (and perhaps misleadingly) entitled "Ageless Bodies." It argues that behind the quest to improve the human life span is an insatiable quest for immortality. The chapter closes with a series of somber rhetorical questions.

> If we go with the grain of our desires and pursue indefinite prolongation and ageless bodies for ourselves, will we improve the parts and heighten the present, but only at the cost of losing the coherence of an ordered and integrated whole? Might we be cheating ourselves by departing from the contour and constraint of natural life (our frailty and finitude), which serve as a lens for a larger vision that might give all of life coherence and sustaining significance?[7]

These sentiments profoundly echo those of the council's former chair, Leon Kass. Writing recently in a conservative journal, Kass goes on record as questioning the value of biomedical efforts at life extension and new initiatives that would alter the human age profile to make possible a longer period of extended, healthy activity at the end of life. He asks, "Does not the human world rather benefit from the cycle of birth, death, and renewal of human life, sad as the death of each person may be to those who love him? If so, is mortality not absurd, but a blessed necessity?"[8]

Can you imagine what would have happened if such questions had been raised by a leading healthcare figure a century ago, when nearly a third of young people died of infectious disease and when sixty years normally marked the end of life? Would we even have seen the revolutions in public health, vaccine, and antibiotic research that have transformed the life experience of millions of people? Can you imagine those questions being put today to a family that has just lost a parent to breast or testicular cancer—or to couples who risk losing a child through a genetic disease? What are we to say to the person just entering his sixties who, anticipating productive years ahead, is diagnosed with Lou Gehrig's disease (ALS) and faces a short and agonizing end of life course?[9] None of these people would regard mortality as a "blessed necessity." Fortunately few people in biomedical research or research policy have taken Kass's or his council's counsels of despair seriously.

The Issue of Germline Genetic Enhancement

I am not going to address here all the concerns that motivate this spirit of retreat. My own record of publication on some of the key new areas of biomedical research—the expansion of the availability of genetic testing; human embryo research; stem cells and therapeutic cloning—is readily available.[10] Instead, I want to focus on one representative area and case: the issue of *germline genetic enhancement.* Should we ever consider altering the human genome in ways that can be inherited by future generations? Should we do so, not just for the purpose of eradicating serious disorders, but also to increase human powers *above and beyond* the level of what is normal for our species?

I choose this topic precisely because it is so controversial. Even many of those who support research in genetics aimed at new strategies for treating disease tend to shrink before proposals for human genetic enhancement. Nevertheless, I am of the opinion that to some extent treatment and enhancement go hand in hand. We cannot have a vigorous research program in genetic medicine that does not also point the way to and facilitate the possibility of genetic enhancements. Furthermore, some enhancements will be a valued part of our future, as they have been a valued part of our past. With the geneticist Hubert Markl,[11] I believe that human beings will increasingly shape our evolution and that it is morally appropriate that we do so.

My aim therefore, is to begin to make the case for human germline enhancement, for the moral permissibility of at least some humanly chosen and inheritable improvements in the biology of our species. It is important that I begin with some definitions. First, there is the distinction between somatic cell and germline genetic interventions. Second, there is the distinction between therapy and enhancement.

Somatic cell gene transfer seeks to alter the bodily cells of an existing individual, usually for the therapeutic purpose of remedying a disease. An example is the trials underway at the Necker hospital in France on X-linked severe combined immunodeficiency disorder (X-SCID).[12] Children, usually boys, who are born with this disease have impaired immune system cells and cannot ward off infections or cancers. The youngsters in this experiment had already failed on the only available alternative treatment, matched bone marrow transplant. In the experiment, some of the bone marrow was removed from each child's body and infected with a virus capable of carrying corrective genetic material into his cells. The genetically transformed cells were then injected back into his body. Because this therapy is applied to bodily and not to reproductive cells, the changes cannot be transmitted to offspring. If one of the boys in this experiment survives to reproductive age and has a son that is born with the disease, that child will have to undergo therapy on his own.

Germline gene therapy, by contrast, affects the reproductive cells. This can be a result of treatment done very early during embryonic or fetal development, when any treated cells are likely to form part of the reproductive system. Or it can result from gene therapy administered to the sex cells of parents. For example, it might be possible in the future to alter the sperm or eggs of X-SCID carriers so that none of the children born to them would suffer from the disease. Largely because of safety concerns and the fear of inadvertently introducing inherited genetic defects into the human population, there is a consensus, supported by most national regulations, that germline gene therapy should not be undertaken at this time. I'll shortly argue that this distinction between somatic and germline therapies will fade in time and that someday in the future germline therapies will likely become a standard part of medical practice.

The second distinction is between therapy and enhancement. In *Beyond Therapy*, the President's Council on Bioethics offers the following definitions as a "useful starting place" for understanding the distinction:

> "Therapy" . . . is the use of biotechnical power to treat individuals with known diseases, disabilities, or impairments, in an attempt to restore them to a normal state of health and fitness. "Enhancement," by contrast, is the directed use of biotechnical power to alter, by direct intervention, not disease processes but the "normal" workings of the human body and psyche, to augment or improve their native capacities and performances.[13]

As you can see, "normality" plays a key role in these definitions. Therapy and treatment aim at relieving or curing known diseases, disabilities, and impairments, which are thought of as involving abnormal bodily states that cause suffering. Our X-SCID youngsters have a genetic disease in this sense—an inborn metabolic insufficiency that threatens their life—and their treatment is appropriately labeled gene therapy. Similarly, medical interventions to help someone with a spinal cord injury, visual impairment, or clinical depression are all therapies in this sense because they address conditions, which, though prevalent to varying degrees, involve impairments of normal functioning. Because therapies seek to relieve suffering caused by abnormal or subnormal bodily states, there is nearly a universal consensus that they are ethically permissible and even required. Indeed, for a treatment course to become eligible for insurance reimbursement in most jurisdictions, it must first be determined to be a recognized disease condition or disability.

By definition, enhancements go beyond the normal. It may be extremely galling to a dedicated young basketball player and his avid fan parents that his likely five-foot-ten-inch mature height will render him too short for success in professional play. But the use of human growth hormone (HGH) to increase his height is an enhancement, not a therapy. (In contrast it is an established therapy to administer human growth hormone to a youngster at least 2.25 standard deviations below the norm for height at his age.)[14] Because enhancements presume that the treated individual is already normal, they are almost never covered by medical insurance. Standards of care in the relevant medical specialties usually discourage professionals from complying with such enhancement requests.

Recently, research on the genetics of muscle tissues has introduced the possibility of using gene therapy to significantly improve athletic performance.[15] Articles in popular magazines speak of "gene doping." Like all enhancements, these possibilities are regarded as ethically questionable, but the questions increase exponentially if such enhancements are made inheritable. Germline genetic enhancements raise the specter of "designer babies" and are widely regarded as one of the gravest dangers of emerging applications of genetic science.

The Moral Gravity of These Distinctions

What are we to make of these distinctions? Is germline gene therapy really as ethically questionable as the critics contend? Are genetic enhancements properly branded a dangerous misuse of genetic science? Are inheritable (germline) enhancements utterly beyond the ethical pale?

Let me begin with the somatic cell versus germline distinction. It is supported by several important considerations, but on closer inspection, they are less compelling. Indeed, making appeal to these same considerations, one could argue for the moral preferability of some germline interventions to somatic cell treatments.

A leading concern in all genetic interventions is safety. At present we know far too little about the complexity of the human genome to warrant most germline therapy. As the President's Council on Bioethics reminds us, "The human body and mind, highly complex and delicately balanced as a result of eons of gradual and exacting evolution, are almost certainly at risk from any ill-considered attempt at 'improvement.'"[16] The French X-SCID research I mentioned earlier, which involves only somatic cell therapy, illustrates the problem. At least three of the eleven youngsters involved in the research have come down with leukemia and one has died.[17] This apparently happened because the viral vector used to carry the corrected gene sequence into their blood cells inadvertently activated a cancer gene in one or more of the thousands of cells that were infected with it. All of the children in the experiment may eventually suffer leukemia and some may die from it, but at least this is not an error that any of the children are likely to convey to their offspring. In contrast, a similar error in the gonadal cells of an embryo or fetus would create a new genetically induced form of leukemia that could be inherited by that person's descendants.

This would seem to be sufficient reason never to permit germline interventions. But there are counterarguments. For one thing, new technologies are being developed that allow for precise targeting of gene modifications in gametes or embryonic cells. These techniques, some of which are still years away from human application, also permit characterization of the resulting cells and verification of insertional accuracy before a pregnancy is begun or an embryo transferred to a womb for development.[18] Other techniques, including the use of artificial chromosomes or controllable antibody cells are also under study.[19] This raises the following ethical question: is it better to require people in generation after generation to undergo somatic cell interventions, with the risk, as in our X-SCID case, that many bodily cells may be inadvertently damaged, or to get rid of the disease-causing gene sequence once and for all with verifiable precision at the gamete or embryo level? It is too early to answer this question, but the answer will grow clearer in the future and the safety argument may eventually tip in favor of germline interventions. For my part, I look forward to the day when couples at risk of having children with such diseases as X-SCID, cystic fibrosis, or sickle cell disease or inherited forms of cancer can undergo very simple (possibly preconception) procedures to eliminate these worries from their future. Indeed, I hope that a century from now, young people will look back upon the standard Mendelian inherited diseases with as much bewilderment as the current generation now looks back on tuberculosis, one of the leading causes of death in 1900.

The distinction between therapy and enhancement is also far less neat than its defenders claim. Consider: between therapy, which remedies disease by restoring people to a normal condition, and enhancement, which aims to improve people over and above the species' normal, there is the intermediate position known as *disease prevention*. Like enhancement, prevention usually involves making people better than normal, but far from being morally questionable, it is among medicine's most laudatory goals.

Almost everyone reading this is an enhanced human being. During infancy, you were routinely administered a set of vaccines that boosted your immune systems well beyond what is "normal" for the human species. Historically, being "normal" meant having a high degree of susceptibility to measles, chicken pox, or whooping cough. If you were exposed to the polio virus, you risked coming down with a crippling disease. Our vaccination initiatives have changed biological human nature and, in the developed world at least, these ancient vulnerabilities no longer worry us.

No one ever morally questions the kind of enhancement that this intermediate possibility, disease prevention, involves. Is there any reason to believe that genetic disease prevention measures should be regarded differently?

To answer my question, let me draw on an example offered by the Princeton biologist Lee Silver.[20] He points out that we now know that 1 percent of people of West European ancestry carry a gene that provides complete resistance to infection by HIV, the AIDS-causing virus.[21] Imagine, now, that we could develop a single-application gene therapy suppository that, when used by an HIV-infected woman at the time of intercourse, would guarantee that any child conceived by her would have the AIDS-resistant gene. Not only would this child be spared the risk of HIV transmission from its mother, but during its lifetime it would be free from worries of AIDS infection. Since children all over the world live under the threat of AIDS, and since even antiretroviral medicines, where they are available, carry significant side effects, an anti-AIDS gene vaccine of this sort would be a godsend. But it would involve germline, genetic enhancement. Would it be morally wrong? The answer is far from obvious. True, there would be many safety questions that would have to be answered before approval of such a vaccine could be recommended. We would certainly have to better understand the health conditions of the natural human controls who currently have the AIDS-resistant gene. But keeping in mind that even established vaccine programs are never risk free, it is very likely that an AIDS vaccine of this sort could pass ethical muster and even meet with an enthusiastic reception.

The Hard Case: Enhancement Unrelated to Therapy or Prevention

If we accept that genetic enhancement in the form of disease prevention can be ethical, we are left with one category of enhancement that continues to trouble many people and around which most of the debate swirls. This is enhancement aimed at producing children that are better than normal in ways unrelated to disease prevention. Goals like increased height, improved memory function, superior athletic performance or musical abilities are often imagined to be on the "shopping list" of parents seeking to give their child a leg up in life.

Those opposed to enhancements of this sort typically raise a variety of objections. Some are more substantial than others. I want to briefly discuss six of the leading concerns that I have found in the literature on this topic. Taken together, I agree that these concerns recommend our taking a far more cautionary approach to non-disease-related improvements than to other germline measures aimed at eliminating or preventing disease. But they do not absolutely rule out all such enhancements, and they even suggest some criteria for identifying those enhancements that may well become common practice in a generation or two.

(1) *Safety*. This is the first concern. As I have said, all meddling with the human genome is risky. This has become especially clear following the discovery, in the wake of completion of the human genome project, that human beings have far fewer genes than was expected: 30,000 compared to 100,000. In order to explain the enormous complexity of human neurobiology alone, genomic scientists now theorize that the genome is highly modular in nature, with many component parts spread physically across the genome allowing different combinatorial events. As Venter et al. observe in their paper accompanying the publication of the human genome sequence, "A single gene may give rise to multiple transcripts and thus multiple distinct proteins with multiple functions, by means of alternative splicing and alternative transcription initiation and termination sites."[22]

This modular architecture and the established reality of gene pleiotropy (the tendency of a single gene to perform more than one function) tell us that even seemingly benign interventions may trigger a cascade of unforeseen effects. It is reasonable to run such risks, of course, when the goal is to cure or prevent serious disease, but it is far less reasonable when the goal is merely to confer an added benefit. To my mind, this safety consideration is the foremost objection to genetic enhancement. Nevertheless, it is not an overwhelming one. The progress of science will certainly help us identify changes in the genome that do not affect other systems. Natural human variations can also be used as testing grounds for future modifications. For example, animal research already provides some evidence that only a relatively few genes may be implicated in increased life span.[23] If such genes can be identified in long-lived individuals, we may well see the day when parents can reasonably, and safely, pick this characteristic for a child.

(2) *Justice.* The prospect of parents selecting a suite of characteristics in their offspring triggers a host of further ethical concerns. A second one that has received much attention in the enhancement literature concerns social justice. If financially well-off people can purchase genetic enhancements for their children, won't the rich not only get richer but also biologically better and, hence, more prone to becoming still richer? If money can buy improved academic performance, athletic ability, or even just greater height and a better appearance, won't genetics contribute to an increasing divide between society's haves and have-nots? Some fear the emergence in the future of a "geneobility," and perhaps in the distant future, the divergence of the human species into two or more separate and no-longer interbreeding species, as in H.G. Wells's novel *The Time Machine*, where humanity has been replaced by two antagonistic races, the docile Eloi and predatory Morlocks.

These are all legitimate concerns, and they will require thoughtful attention in our genetic future. But they do not point unequivocally toward prohibition. Just the opposite. They may require that opportunities for enhancement be widely disseminated. This might mean the inclusion of a basic suite of enhancements in the package of medical benefits offered by society and made available to everyone. Just as we now believe that all deserve access to the available roster of vaccinations, so we may conclude that a guaranteed minimum life span, stature, or even IQ is a social right that should be afforded to every citizen at birth.[24]

What is interesting here is that enhancement genetics, which can, in one respect, be seen as a threat to social justice, may actually furnish the tools for moving forward toward a more just society, one in which the natural lottery of genetic fate is partially modified by deliberate efforts at human control. Writing over thirty years ago in his landmark book, *A Theory of Justice*, my own teacher, the philosopher John Rawls, held out a vision of a just society that includes efforts by people to "to insure for their descendants the best genetic endowment."[25] In my more perverse moods, I sometimes wonder whether at least some of the emotional energy directed against genetic enhancement doesn't come from people who, having already done well in nature's arbitrary genetic lottery, are uncomfortable with the thought that other, less "naturally gifted" people may someday be able to crowd their way into the circle of the poised, the smart, and the beautiful. In any case, this thought reminds us that while enhancement genetics can easily become a new opportunity for human greed, properly handled, it can also contribute to our moral purposes as a species.[26]

(3) *Futile quests for positional advantage.* A third concern is the fear that in a competitive society like our own, children themselves will become the victims of a competitive and ultimately self-destructive "arms race" in genetic enhancements. Take height as an example. It is well established that in both males and females, added height is positively correlated with career success. For each additional inch in stature, the income of adult males goes up by nearly 1 percent or about $800.[27] We can imagine, therefore, that if genetic means existed to enhance the stature of their children, parents might feel pressure to do so, with the absurd result that the relative advantage of height would decline as more and more children are born tall. In the end, children would be exposed to the dangers of genetic meddling for little or no purpose. In the search for private positional advantage, parents would create a kind of "tragedy of the commons" whereby everyone, including their own offspring, would suffer. Related to this is the worry that parents will become entranced by "harmful conceptions of normality" and that monocultures of movie star look-alikes will slowly replace natural human diversity.[28]

These are real concerns. Nevertheless, those who voice them exaggerate the risks. Parents in every generation want the best for their children, including access to the opportunities and resources needed to accomplish the child's life dreams. How many parents, informed of the risks accompanying more complex interventions, will needlessly risk their child's health to achieve competitive advantage in schooling or business? Nor will reasonably wise parents seek to impose their own visions of excellence on a child when there is good reason to believe that the child will have its own mind (although a small number of troubled parents may do this, just as they now do without genetic tools). Instead, I believe, most parents will avail themselves of a suite of modest enhancements, small improvements over their own perceived weaknesses, perhaps, to ensure that their kids have a good start in life. Beyond this, they will leave it to the child's own pluck and personality to identify and fulfill his or her dream.

A further observation here is that while most parents want their children to have some increased opportunities in life, they don't want children who are radically different from themselves. As John Lachs reminds us,

> at least a part of the charm of children is that they are like us; if they were radically better, we could not have the cozy sense of their being ours. Parents dote on children who look and act like them. Being prettier or a better athlete is welcome if the increment does not interfere with recognizing ourselves in the next generation. But purging our offspring of our imperfections makes them at most admirable strangers, not our beloved own. It makes it hard for us to believe that we had anything to do with begetting or raising such exquisite little people.[29]

(4) *Deformations of parenting.* A fourth concern focuses less on the child than the parents. The practice of genetically enhancing our children, it is argued, will somehow lead to a deformation of parenting. Gilbert Meilaender, a member of the President's Council on Bioethics, whose sentiments resonate throughout several key council documents, expresses this concern in an essay entitled "In Search of Wisdom: Bioethics and the Character of Human Life."[30] Meilaender quotes at length from Galway Kinnell's poem, "After Making Love We Hear Footsteps," which celebrates the arrival of a young child in the parents' bed. Flopping down between the couple, he hugs his parents and "snuggles himself to sleep, his face gleaming with satisfaction at being this very child."

In this moment, Meilaender says, the poet offers us the image of the child "as a gift that is the fruition not of an act of rational will but an act of love." This image, he maintains, "can be contrasted with an image of the child as the parents' project or product . . . we undertake to satisfy our purposes and make our life complete." According to Meilaender, such a project- or product-oriented approach to parenting, an approach he believes is inevitably associated with prenatal genetic interventions, undermines the openness that is essential to respecting the child's own development as a free and unique human being.

To his credit, Milaender acknowledges that there are limits to parents' "openness" to the "gift" that a child represents. A few paragraphs further on, he observes, "there is no simple recipe for making decisions. Parents must indeed exercise reason and will to shape their children's lives. They do not and should not simply accept as given whatever disabilities, sufferings, or (even just) disappointments come their children's way." But he then hastens to add, "Still, as every child realizes at some point, the conscientious parent's effort to nurture and enhance can be crushing."

Of course it can be. Crushing soccer moms and football dads exist all around us. I have seen toddlers decked out in T-shirts labeled "Dartmouth Class of 2024." But despite the enthusiastic excesses of a few, most responsible parents wage a constant struggle to balance their efforts to mold and shape their child's course with a sense that a child must be left free enough to find his or her own way in the world. In the words of ethicist William Ruddick, parents always play a dual role as "guardians and gardeners."[31] Parents' love is unconditional, but their actual responsibilities to the child require repeated interventions to shape, mold, and direct in ways that will most enrich the child's life.

(5) *Absence of Consent.* What about the children themselves? Should they have a say in all this? A fifth concern often voiced in this context is that genetic manipulations, unlike many other things that parents do to children, proceed entirely without the child's consent. If I select a religious community for my child, or an educational path, the child can participate in some way in the decision process. Even young children, though not capable of full consent,

can express *assent* to decisions made on their behalf or withhold assent by resisting.[32] Although it can be very difficult, later in life they can try to reverse educational or religious choices that parents have made for them. But where germline genetic changes are involved, no such assent or resistance is possible. One either has or does not have the gene.

In fact this last point is not quite correct. Research by the geneticist Mario Capecchi involving CRE recombinase opens the possibility that some gene insertions could be turned on or off at any time in the future and in any subsequent generation by the administration of medications that either delete or activate the added gene sequence in all the cells of a person's body.[33] This has implications for the consent issue. It would allow musically inclined parents to confer upon a child an ability such as perfect pitch, and the child could choose to switch this on (or off) as she wished. The same technology applied to all germline gene therapies would allow us to put an end to interventions that, over time, proved undesirable.

If technology might help here with the consent problem, so, too, can wise ethical choice. Parents can give children enhanced abilities that are neutral or positive for many life choices. Improved visual acuity is an example. Even now, some top athletes like Tiger Woods undergo laser eye surgery procedures to achieve better than twenty-twenty vision. Genetically endowed enhancements like this do not impair a child's freedom. Instrumental to the pursuit of a wide variety of career plans, they are among the things that parents might allowably choose to improve their child's range of options. They are the kind of permanent changes we might make in the genome with the confidence that future generations will not regret our choices.

(6) *Character and authenticity.* A sixth and final concern often voiced in this regard has to do with issues of character and authenticity. It frequently finds expression in the realm of athletics, where technological interventions aimed at winning are sometimes said to undermine the exercise of genuine human excellence. More generally, it is fear that by intervening to manipulate the biological bases of various human skills and talents, we will somehow lose what is human about the very process of pursuing these goals. Laboratory accomplishments will replace effort, dedication, and genuine human achievement.

There are so many different worries here that it is hard to separate them out, much less do justice to them all in the space of a few paragraphs. It helps to realize that we hear this concern most often expressed in connection with the use of performance-enhancing drugs in athletics. Here, the judgment that such "artificial" measures undercut effort and trivialize the display of human excellence makes good sense. Athletes who rely on injectable steroids or blood doping are not only cheating and taking a shortcut to success; they are risking their own health. Indeed, we take steps to close this shortcut (and declare it a form of cheating) because it is a kind of behavior whose risks make it something we want to discourage. This risk concern is magnified by the power of drug interventions to affect outcomes. In all competitive endeavors, we are understandably wary of interventions that reduce the full range of competitive activity to the purchase of a tool, and a dangerous tool at that.

Thus, our opposition to the use of drugs in athletics is not the result of the single judgment that pharmacology is artificial and therefore inimical to athletic effort or excellence. Rather, it is the conclusion of a complex reasoning process involving several different and intersecting considerations. That we do not regard the use of laboratory-developed artificial aids per se as undermining athletic effort or achievement is shown by the fact that we fully permit the use of graphite tennis racquets, fiberglass vaulting poles, and engineered running shoes or swim suits. We also routinely allow athletes to undertake specialized diets to improve performance. We don't ban such things partly because there is little need to do so—most are not dangerous to health—and partly because we judge excellence to be the outcome of many factors working in concert: one's native endowment, the efforts one makes to develop it, and the modest aids or boosters one can use to bring all this to the highest level of performance.

Genetic interventions in sports and elsewhere will have to be evaluated in the same way. Risks, once again, will play a major role in our thinking, and we will steer away from interventions that come with a high price in terms of human health. We will be wary of genetic enhancements that can overwhelm the many other excellences and skills that go into competition. And we will always worry about licensing parents to coerce children into a parental lifestyle through early genetic choices. But I doubt that issues of character or the authenticity of performance by themselves will amount to much. Already in athletics one's competitive prospects are very much shaped by one's genes and no one maintains this purely inherited factor undermines character or achievement. Instead, achievement is the product of one's genetic start and the things one does with it. This suggests that harmless and otherwise unobjectionable genetic enhancements—improved eyesight and better muscle tone are examples—will not only be integrated into our conception of excellence but will routinely become part of the culture of families devoted to athletics.

If we imagine genetic aids being integrated into our notions of excellence in an area like sports, they are likely to become even more so in areas where high levels of performance have not just competitive value but intrinsic social worth. Professionals in demanding lines of work already use pharmacological enhancements to improve mental alertness.[34] I cannot imagine that it would reduce our esteem for the talents of a world-class brain surgeon, or make us less willing to hire her, to learn that her talents were partly owed to genetically enhanced fine motor skills.

Our Posthuman Future?

I have only begun to make what is an enormously complex case. There are many reasons to worry about interventions that alter the human genome. Safety concerns are foremost. They grow in direct proportion to the heritability of the changes and in inverse proportion to the intervention's relationship to curing or preventing disease. Therapy justifies substantial risks, whereas enhancements must prove their safety. For non-disease-related enhancements, something like the precautionary principle would thrust the burden of proof for each intervention on the promoters of change, not its opponents.[35] Enhancements also raise questions of social justice and threaten to unleash fruitless quests for positional advantage. As the critics note, they can invite well-meaning parents to exercise too much control over a child's life.

Other concerns are less substantial. The focus on character and authenticity, I believe, fades away unless it can be linked to interventions that are harmful or otherwise counterproductive. Human excellence is already the product of too many arbitrary factors, not least of all natural genetics, to argue for some pristine model of pure self-achievement.

One concern that need not worry us at all at this time is the fear that genetic enhancements will somehow take us into a "posthuman future" where fundamental human rights are placed in danger. This concern has been voiced by several of our cultural critics of biotechnology, notably Francis Fukuyama, whose fears appear to have found an enthusiastic hearing among the other anti-biotechnology members of the President's Council on Bioethics.

Fukuyama calls for immediate regulation of the whole sphere of genetics on the grounds that this technology threatens to "disrupt either the unity or the continuity of human nature, and thereby the human rights that are based on it."[36] He seems to believe that even the slightest changes in our biological constitution will somehow dissolve our age-old commitments to fairness, justice, and compassion. He evokes an impending world of such dramatic biological change that some will be bred with metaphorical saddles on their backs and others with boots and spurs to ride them.[37]

But how can any of the modest changes parents are likely to effect have such consequences in the foreseeable future? Consider: in the space of a century, we have already dramatically extended the human life span by a third; we have freed much of our species (sadly, not enough of it) from vulnerability to ancient scourges; we have transformed human sexuality through contraception and Assisted Reproductive Technologies so that the life of a modern woman would be almost unrecognizable and shocking to a woman or man of 1900; and we have even altered the physical appearance of the species by eliminating many of the minor blemishes, from cleft palate to bad teeth, that routinely affected our forebears. Has any of this disrupted human rights? Just the opposite. One could say that access to better and more evenly distributed conditions of biological well-being have actually accentuated the importance of human rights. Individuals who come to expect the fulfillment of their desires, who take health and long life as a fundamental right and not an accident of birth, also expect better treatment in all the other areas of their life. This partly explains the demands by many seniors today for an expanding and improved system of Medicare coverage.

The calls to slow and regulate biotechnology are not only unnecessary, but also dangerous. If researchers are allowed to follow their interests and are free to find support for their work without arbitrary political intervention, a century from now we will see at least the same degree of progress that has marked the past one hundred years. The average human life span will have been extended another 30 percent. With conditions like Parkinson's and Alzheimer's disease, osteoporosis and rheumatic disease brought to heel, the later years of life will be much freer of the many forms of morbidity and dependency that currently mar it. For those in midlife, many of the causes of premature death will be better understood and eliminated. The eradication of many genetic diseases and other inherited predispositions to suffering will give most young people a good start in life. Removing the many blemishes that mar self-esteem, from poor teeth or a bad complexion to a propensity to eating disorders and anxiety, will unleash a new burst of human self-confidence.

All this can happen if we permit scientists to follow their dreams. We should strive, within a reasonable framework of human subject protections, to let the cutting edge science of each generation exercise its potential. In our generation genetics is one such area of science. We are close to forgetting that excellent science is a delicate flower that does not easily survive abrupt legal moratoria on research, onerous and unreasonable regulations, or a hostile cultural environment that drives away young researchers or investors. We are already seeing these adverse impacts in the areas of stem cell, therapeutic cloning, and gene therapy research, where the United States is falling behind countries in Asia.[38] We must not succumb to the complacency that says that the human organism has achieved its optimum state and we can now relax our efforts at health improvement. Above all, we must not lose our capacity to dream.

Notes

This chapter is dedicated to the memory of my nephew, Clifford Naer, who recently died suddenly at age 44.

1. My source for all these statistics is the Center for Disease Control, "Achievements in Public Health, 1900–1999: Healthier Mothers and Babies," *MMWR Weekly* 48:38 (October 1, 1999): 849–858, http://www.cdc.gov/epo/mmwr/preview/mmwrhtml/mm4838a2.htm

2. Infoplease, "Life Expectancy by Age, 1850–2001," http://www.infoplease.com/ipa/A0005140.html.

3. Infoplease, "Life Expectancy by Age, 1850–2001," http://www.infoplease.com/ipa/A0005140.html.

4. For white males at birth and ten years of age, the figures are seventy-five and sixty-six; for non-white females, seventy-six and sixty-seven; and for non-white males sixty-nine and sixty.

5. Peter Sloterdijk, "The Operable Man," http://www.goethe.de/uk/bos/englisch/Programm/archiv/2000/enpslot200.htm.

6. President's Council on Bioethics, *Beyond Therapy: Biotechnology and the Pursuit of Happiness* (New York: ReganBooks, HarperCollins, 2003), available at www.bioethics.gov.

7. Ibid., 200.

8. Leon R. Kass, "Human Frailty and Human Dignity," *The New Atlantis* 7 (Fall 2004 / Winter 2005): 111.

9. John Schwartz, "Longtime Expert on A.L.S. Now Knows It All Too Well," *New York Times,* February 22, 2005: Final, F1.

10. My curriculum vitae is available online at: http://www.dartmouth.edu/~ethics/rg_cv.html.

11. Hubert Markl, "Man's Place in Nature: Evolutionary Past and Genomic Future," *Journal of Molecular Biology* 319 (2002): 869–876.

12. Salima H. Hacein-Bey-Abina et al., "Sustained Correction of X-Linked Severe Combined Immunodeficiency by ex Vivo Gene Therapy," *New England Journal of Medicine* 346 (April 18, 2002): 1185–1193.

13. *Beyond Therapy*, 13. My own approach to the definition of disease and enhancement is greatly informed by the work of Bernard Gert, Charles M. Culver, and K. Danner in their *Bioethics: A Return to Fundamentals* (New York: Oxford University Press, 1997), chapter 5; and Charles M. Culver, "The Concept of Genetic Malady," *Morality and the New Genetics*, ed. Bernard Gert et al. (Boston: Jones and Bartlett, 1996). The work of Gert et al. also influences Eric Juengst. See his "Can Enhancement Be Distinguished from Prevention in Genetic Medicine," *Journal of Medicine and Philosophy* 22 (1997): 125–142; and "What Does 'Enhancement' Mean?" in *Enhancing Human Traits: Ethical and Social Implications*, ed. Erik Parens (Washington, D.C.: Georgetown University Press, 1998).

14. See Dov Fox, "Human Growth Hormone: The Measure of Man," *The New Atlantis* (Fall 2004 / Winter 2005): 83.

15. Jesper L. Andersen, Peter Schjerling and Bengt Saltin, "Muscle, Genes and Athletic Performance," *Scientific American* 283:3 (September 2000): 48–55.

16. *Beyond Therapy*, 287.

17. "French Gene Therapy Trial Suspended Again," *Bionews* 293 (January 24–30, 2005), http://www.bionews.org.uk/new.lasso?storyid=2427.

18. William M. Rideout III et al., "Correction of a Genetic Defect by Nuclear Transplantation and Combined Cell and Gene Therapy," *Cell* 109 (April 5, 2002): 17–27.

19. Mario Capecchi, "Human Germline Gene Therapy: How and Why," in *Engineering the Human Germline*, ed. Gregory Stock and John Campbell (New York: Oxford University Press, 2000), 31–48; Katherine High, "Anemia and Gene Therapy—A Matter of Control," *New England Journal of Medicine* 352:11 (March 17, 2005): 1146–1147.

20. Lee M. Silver, "How Reprogenetics Will Transform the American Family," *Hofstra Law Review* 27:3 (Spring 1999): 649–658.

21. More recent research indicates that the correct number is 13 percent. See Michael Balter, "Are Humans Still Evolving?" *Science* 309 (2005): 234–237.

22. J. C. Venter et al., "The Sequence of the Human Genome," *Science* 291 (2001): 1304–1351.

23. See S. Austad, "Adding Years to Life: Current Knowledge and Future Prospects," presentation at the December 2002 meeting of the President's Council on Bioethics. Transcript available at: http://bioethicsprint.bioethics.gov/transcripts/dec02/session1.html.

24. See Maxwell Mehlman and Jeffrey Botkin, *Access to the Genome: The Challenge of Equality* (Washington: Georgetown University Press, 1997); and Maxwell Mehlman, *Wondergenes: Genetic Enhancement and the Future of Society* (Bloomington: Indiana University Press, 2003). While Mehlman is willing to permit some access to somatic cell enhancements, due to considerations of justice, he appears to be utterly opposed to inheritable (germline) enhancements. See *Wondergenes*, 183–191.

25. See John Rawls, *A Theory of Justice*, (Cambridge, Mass.: Harvard University Press, 1971), 108.

26. See Norman Daniels, "Normal Functioning and the Treatment-Enhancement Distinction," *Cambridge Quarterly of Healthcare Ethics* 9 (2000): 309–322.

27. Dov Fox, "Human Growth Hormone: The Measure of Man," 76. See also, Stuart Buck, "The Buck Stops Here: Height and Income," *Yahoo News*, October 17, 2003, http://stuartbuck.blogspot.com/2003/10/height-and-income.html.

28. Erik Parens, "Is Better Always Good: The Enhancement Project," in *Enhancing Human Traits: Ethical and Social Implications*, ed. Erik Parens (Hastings Center Studies in Ethics, Washington, D.C.: Georgetown University Press, 1998), 19.

29. See John Lachs, "Grand Dreams of Perfect People," *Cambridge Quarterly of Healthcare Ethics*. 9 (2000): 323.

30. See Gilbert Meilaender, "In Search of Wisdom: Bioethics and the Character of Human Life." Paper presented and discussed at the January 2002 meeting of the President's Council on Bioethics, available at: http://www.bioethics.gov/background/meilaenderpaper.html.

31. See William Ruddick, "Parenthood: Three Concepts and a Principle," in *Family Values: Issues in Ethics, Society and the Family*, ed. Laurence D. Houlgate (Belmont,

Calif.: Wadsworth 1988), http://www.nyu.edu/gsas/dept/philo/faculty/ruddick/papers/parenthood.html.

32. Assent by children seven years or older to research in which they are subjects is a formal requirement of human subjects research regulations. See Title 45 Code of Federal Regulations, 46, Subpart D.

33. Mario R. Capecchi, "Human Germline Gene Therapy: How and Why," in *Engineering the Human Germline*, ed. Gregory Stock and J. Campbell (New York: Oxford University Press, 2000), 31–42.

34. Anjan Chatterjee, "Cosmetic Neurology: The Controversy over Enhancing Movement, Mentation, and Mood," *Neurology* 63 (September 2004): 968–974.

35. Edward Soule, "Assessing the Precautionary Principle," *Public Affairs Quarterly* 14:4 (October 2000): 309–328.

36. Francis Fukuyama, *Our Posthuman Future: Consequences of the Biotechnology Revolution* (New York: Farrar, Straus and Giroux, 2002).

37. Ibid., 10, 154.

38. See Dennis Normile and Charles C. Mann, "Cell Biology: Asia Jockeys for Stem Cell Lead," *Science* 307:5710 (February 4, 2005): 660–664; and Andrew Pollack, "Cancer Therapy Dropped in U.S. Is Revived in China," *New York Times*, February 25, 2005, http://www.nytimes.com/2005/02/25/business/worldbusiness/25biotech.html.

FOUR

Biotechnology in a World of Spiritual Beliefs

LEE M. SILVER

Biotechnology provides the greatest hope for alleviating human suffering and, simultaneously, sustaining a vibrant biosphere. Human civilization was founded on the ability to control and manipulate genes in other organisms, including plants, animals, and microbes; neither cows nor corn existed until human ancestors invented them. But with its ancient history largely unknown, biotechnology appears as the most contentious of modern inventions because it challenges overt, covert, or subliminal Western beliefs in a "Supreme Being" or a preordained "Master Plan." From conservative Christians who view embryo cloning as a violation of God's right to create each individual human soul, to New Age secularists who view plant engineering as an assault on the spirit of Mother Nature, opponents of biotechnology are driven by faith in a higher or deeper spiritual authority who demands our allegiance. In contrast, the fluid spiritual traditions of Asian countries—where souls are eternal, self-evolving, and not beholden to an external master—allow a more ready acceptance of both embryo research and genetically modified plants. Because of fundamental differences in Western and Eastern spiritual traditions, Korea, China, Singapore, and India are now poised to take the lead in biotechnological advancement.

The Agricultural Revolution and Human Civilization

The long history, legacy, and pervasive impact of biotechnology on humankind and the biosphere as a whole is not fully appreciated even by most well-educated people. Although it is commonly thought to be an invention of twentieth- and twenty-first-century scientists working inside brightly lit high-tech labs, biotechnology was first developed at the end of the last Ice Age, between eight thousand to twelve thousand years ago, at multiple independent locations around the world.[1]

As the Ice Age drew to a close, human populations grew rapidly and spread out across the subtropical and temperate zones of the Americas and Eurasia. Hundreds of species of large game animals were hunted into extinction, and edible vegetation was over-foraged. In the past—when the human footprint on the world was smaller—a tribe could simply get up and move from a nutritionally depleted habitat onto virgin land where natural resources were still plentiful. But once habitable virgin land was exhausted, the entire biosphere was pushed beyond its human-carrying capacity.[2] Any other species would have collapsed under the weight of its own voracious appetite, but human genes had endowed human beings with the capacity to initiate a revolutionary lifestyle change that blew apart the traditional equations of adaptation and survival. Instead of fitting into a natural world as best as they could—like every creature before—the human species *consciously* took control away from Mother Nature and into its own hands through a process we now refer to as the Agricultural Revolution.

The Agricultural Revolution emerged out of the human discovery of *genes*—the invisible abstractions that carry specific characteristics of plants and animals from one generation to the next. Genetic conceptualization allowed people to create novel organisms expressing *domesticated* characteristics built to satisfy both human needs and their newly emerging desires. In Central America, a slender weed named teosinte with a few dispersible hard seeds was transformed into cobs of corn with tightly attached kernels that only come off when we want them to.[3] Corn looks nothing like the teosinte weed it was engineered from. In fact, scientists wouldn't even know they were related without the tools of modern genetic analysis.

The independent discovery of genes allowed people living in the Middle East to transform an entire series of weeds into wheat, peas, chickpeas, and lentils.[4] In South America, shrubs from the poisonous nightshade family (*Solanaceae*) with tuberous roots, spiny branches, and bitter berry-sized fruit were transformed into juicy red tomatoes, potatoes, sweet potatoes, and peppers. In South Asia, the chromosomes of wild inedible weeds from Malaysia and India were combined to create banana trees so overloaded with DNA that their fruit can't produce any seeds and are completely sterile. Propagation of seedless bananas over the subsequent millennia has depended on the human application of *cloning*. And in Central Asia, *grafting* technology was perfected for growing bi-species trees with roots from hearty but inedible crabapple stocks and stems from mutant crabapple trees bred to produce what we've come to know as apples.

Like plants, animals were modified by people to create bio-factories that are ever more efficient in the generation of food and other valued products.

But in addition, animals presented people with opportunities to control another set of characteristics—behavioral instincts. Indeed, animal domestication is defined by genetic modification of behaviors. Wild, human-threatening, and human-fearfulness instincts are eliminated and replaced by tameness, an acceptance or desire to be near humans, and often, other specific human-serving personalities.

Chickens can be traced back to a few dozen Red Jungle Fowl (*Gallus bankiva*) that flew in the forests of northeast India six thousand years ago. By 1400 BC, chickens were used for eggs and meat across Europe, Africa, and Asia. In rural areas of underdeveloped countries like Indonesia, Vietnam, Burma, and Belize, chickens are still raised without fences so they can forage widely during the day and indirectly increase the effective amount of land available to provide nutrition for a peasant family. They return each evening to roost where their eggs are removed and eaten, and where they themselves eventually end up on a dinner plate.

The prize for the most extreme and diversified alterations of a single animal species definitely belongs to the descendants of the East Asian gray wolf.[5] From this one species, people have bred a vast variety of dogs ranging in size and appearance from two-pound pocket Chihuahuas to the 200-pound Saint Bernard. Many dog breeds, unlike wolves or other species, have become genetically hardwired to read subtle nonverbal human communicative cues.[6] But, in addition, different breeds express entirely different, human-selected, instinctive behaviors from herding sheep to pointing, flushing, retrieving, or hunting for game to guarding territory or simply providing empathetic companionship.[7]

In the Middle East, oxen were transformed into docile ten-gallon-a-day milk-producing factories with absurdly large udders; their mammary glands were expanded, the amino acid composition of their milk was altered to better suit human nutrition, and the weaning age of calves was reduced so that humans could use them more quickly to convert grass into milk.[8] Also in the Middle East, a hair-covered goat was bred into domestic sheep that grow unnatural, billowing coats of wool, and wild boars were transformed into pigs that subsist on garbage, breed profusely, mature rapidly and live comfortably in close association with people. As a result, pigs provide more meat for worldwide human consumption than any other animal.[9]

Once an animal was domesticated for one purpose, it made sense to select it simultaneously for additional products (although modern-era farmers have reverted to selecting specialized varieties for one use only). Chickens provide eggs, meat, and feathers for insulation; cows are used for meat, milk, and leather; and sheep give us meat and wool. Finally, after the domestication of plants and animals, ancient biotechnologists discovered and developed the biochemical capacity of the third "kingdom" of living organisms—microbes—to create new products like cheese, vinegar, wine, and soy sauce that satisfied the need to prevent food spoilage, as well as the human desire for novel gustatory and drug-induced mood experiences.

Biotechnology's significance is hard to exaggerate. Its invention represented a fundamental turning point in the story of the human species and provided the gateway to civilization. With the domestication of plants and animals, tribes were no longer limited in size by the nutritional capacity of their natural environment. When an undisturbed plot of woods or brush was converted into a cornfield or rice paddy, the yield of edible vegetation could be multiplied by millions. And if properly managed, the same rice paddy could be regenerated year after year to produce the same high yield. Farmers produced more food than was required for themselves and their families, and the surplus became available for trading to specialty craftsmen for other desired objects or services. Domesticated organisms themselves—cows, pigs, chickens, corn, wheat, rice, and a few dozen other useful biotech creations—flowed along migration and trade routes across the Americas, Europe, Asia, and Africa.[10] Trades became professions that exploded in diversity as tribal settlements grew into villages, villages grew into towns and cities, and cities joined together to become nations with evermore diversified economies and complex technologies.[11] And although no one knew it at the time, within a brief moment of the history of life on earth, the relationship of our species to all others was forever changed.

The Green Revolution and Avoidance of Famine

Throughout the first ten thousand years of the biotech era, the rate of innovation was obviously stupendous but it was still held in check by the very low rate at which genetic alterations—mutations—arise spontaneously. In the first half of the twentieth century, however, a huge technological advance occurred when scientists discovered, by chance, that high-energy radiation and certain mutagenic chemicals could induce mutations at a hundredfold greater rate. By 1967, artificial mutagenic methods had been clearly validated, and an international community of agricultural scientists, representatives of developed countries, and humanitarian aid groups came together to exploit its power for the purpose of improving the lives of subsistence farmers in the least developed countries of the world. The International Rice Research Institute (IRRI) was set up in the Philippines, and the International Maize and Wheat Improvement Center (CIMMYT) was set up in Mexico. Together, these institutes focused their efforts on the three most important crops in the world.

Just as the globally supported effort to develop new crop varieties was getting underway, in 1968, the Stanford ecologist Paul Ehrlich published his bestselling book, *The Population Bomb*.[12] In his opening paragraph, Ehrlich wrote grimly, "The battle to feed all of humanity is over. In the 1970s the world will undergo famines—hundreds of millions of people (including Americans) are going to starve to death." Thirty-seven years later and doomsday has yet to arrive. Why? Ehrlich's main mistake was spiritual rather than scientific. In his heart, he couldn't accept the idea that biotechnology might benefit humanity. What actually happened to food production during the three decades after *The Population Bomb* was released is now called the Green Revolution.

Beneficent publicly supported green revolution biotechnology led to the creation of thousands of varieties of crops with increased disease and pest resistance, increased tolerance to drought and poor soil conditions, and increased nutritional value.[13] In one striking example (out of many), new varieties of rice—which previously could only be grown in one season—can now be planted and harvested multiple times each year. Yields have doubled and costs of production have been slashed as genetically improved seeds were distributed to poor farmers in Indonesia, India, Mexico, and other countries in Asia and Latin America.

Indeed, today, anti-biotech activists knowingly take a stance opposite to the one preached by Ehrlich when they argue, "we can already grow enough food for everyone—starvation is due mostly to the unequal distribution of food, political posturing and the economic power of the wealthy."[14] Indeed, they are correct. Current farmland could produce enough food to feed everyone in the world, but that's only because of biotechnological innovation, a fact that biotech opponents typically ignore. They also refuse to understand the fact that biotechnology has never been *just* about making sure there's enough food to eat. When the costs of producing and consuming food are reduced, more money is available for people to spend elsewhere, which allows them to increase the *quality* of their lives.

Modern Biotechnology Applied to Non-Human Organisms

Until the 1970s, changes in the DNA of an organism could not be controlled by conscious beings. But then a new chapter of biotechnology burst onto the scene as molecular biologists developed increasingly sophisticated methods for precise control over the design and implementation of particular DNA alterations. At first, they learned simply how to move genes from one organism to another. As a poignant example of the power of this first-generation, targeted gene modification technology, a microbe was created that produces human insulin, which provides diabetics with a cheaper, more natural alternative to the pig pancreas insulin they previously required to live a normal life. Microbes that produce many other therapeutic human proteins have also been created in a similar manner by splicing functional human genes into the microbial genomes.

The methods of genetic engineering are now much more sophisticated, allowing the implementation of far more subtle changes than whole gene swapping. The subtlest change of all is the switching of single letters in a sentence of gene code in a predetermined manner; for example, from GCGA-GAGTTC to GCAAGAGTTC. Most spontaneous mutations that occur in nature also cause single-letter changes, but they are entirely random and unpredictable. The difference in efficiency between the "natural" method of genetic modification and the modern biotech approach is well illustrated with an actual example from the history of domesticated pig breeding. Around one thousand years ago, a single piglet was born with a particular single-letter mutation (from a G to an A) in a DNA region that modulates the activity of a gene coding for a particular growth factor.[15] The mutation causes increased gene activity in muscle tissue with a consequent 3–4 percent rise in the meat portion of the animal. At the same time, it reduces gene activity in fat tissue with a consequent 3–4 percent reduction in total fat. Medieval farmers couldn't see the DNA change, of course, but they could detect its subtle effect on producing a meatier, less fatty pig. Present-day DNA analysis tells us that once the mutation appeared, farmers selected animals that possessed it, and animal traders spread it into domesticated pig stocks across the entire world.

Now let us suppose that the single-letter DNA mutation had never occurred, but modern molecular biologists figured out, nonetheless, the advantage it could provide to pork production. A pig breeder would then have two choices for obtaining an animal with the specific DNA change, which could then be bred to produce a whole stock of animals with the same mutation. With the old-fashioned random approach, he would have to breed and test approximately one billion animals (an outrageously expensive undertaking) before he was likely to find a single founder animal for stock production. With modern biotechnology, he would simply create the desired modification in pig embryo cells cultured in the lab, and then develop the appropriately modified embryo into a pig.

With intense commercial interest in both pig and cow farming, the new tools of molecular analysis will surely be used to identify many other single-letter and multiple-letter changes in the pig or cow genomes that could increase an animal's value, provide a healthier product for humans to consume, or reduce the animal's negative impact on the environment. With directed genetic engineering methods, each imagined change could be efficiently implemented in the production of animals with appropriately modified genomes. Each genetic change could also occur by chance, just like the more-meat / less-fat mutation in pigs. So why should the process matter so much to organic food enthusiasts and others if the outcome is the same?

In less than thirty years, the power and accomplishments of modern biotechnology have already been mind-boggling. In the agricultural domain, crops and animals have been modified to provide nutrition and calories with enormously improved efficiency at every stage of production, using processes that are much more friendly to the environment than traditional agriculture.[16] Plants and animals are also being deployed as pharmaceutical factories, and large animals (cows, sheep, pigs) are being engineered to produce humanized blood for transfusions, humanized milk to replace infant formula, and humanized organs for transplantation.[17] But ironically, as molecular biology brings precision and transparency to the actual genetic and cellular modifications that biotechnologists perform, it also shines a brighter spotlight on contended connections between organismal life and spirituality. Not all applications of biotechnology to plants and animals will have benefits that outweigh costs, but each potential idea and targeted implementation can be evaluated on a case-by-case basis, which is more than is required currently for crops derived from randomly mutagenized seeds or new animal breeds.

Stem Cells and Personalized Cell Therapy

The expanding knowledge and tools of molecular and cellular biology have also been applied to the development of methods for regenerating living tissues and organs from a patient's own cells. The most straightforward method, so far, involves the conversion (or *cloning*) of the patient's cells into embryos that can be expanded into an unlimited number of cells that remain in an embryonic state indefinitely. These embryonic cells can then be channeled into the development of brand-new tissues and organs as replacements for body parts that are diseased or not functioning properly. Since therapy is carried out with the patient's own cells, there is no chance of immunological rejection. Until 1998, most scientists had no reason to think about using human embryos in their experiments. The U.S. government refused to provide any money for this kind of research and, in the United States, it was conducted entirely by reproductive biologists working in fertility clinics with private funding. The research served only one purpose—improving the rate at which patient pregnancies could be achieved through in vitro fertilization (IVF) and other reproductive technologies. Then in 1997 and 1998, two independent and unexpected scientific breakthroughs provided the conceptual framework for a biomedical revolution that could change the practice of medicine in rich societies and poor ones alike.

Every cell in an adult human body (with a few exceptions) carries the entire genetic code, or genome, present in the one-cell embryo out of which that body developed. Cells appear differently and function differently not because of a difference in genes but because of a difference in the way those genes are being used. Each cell activates only a subset of the twenty-five thousand genes in the human genome. Some genes are activated in all cells; others are activated in some types of cells, while others still function only in a single cell type. The totality of gene activity expressed in a particular cell is considered that cell's *genetic program*. The genetic program, in turn, is determined by the highly specific attachment of a cell's regulatory proteins to the DNA.

This basic scientific understanding laid the foundation for the cloning experiments brought to fruition, by Ian Wilmut, in 1996.[18] Wilmut and his colleague Keith Campbell took the isolated nucleus of an adult sheep cell and placed it into a sheep's egg whose own nucleus had been removed. Regulatory proteins in the egg cytoplasm entered the newly juxtaposed nucleus and *reprogrammed* the activity of its genes—turning some on and others off—into the overall pattern that defines an embryo cell. Forcing a nucleus to undergo reprogramming in this crude manner was rather inefficient, and in their first experiment with 277 reconstituted cells, Wilmut and Campbell got only a single embryo to develop into a fetus and animal with genes equivalent to those of a sheep that had lived previously. The animal was Dolly, who was presented to the world in February 1997. Since her birth, tens of thousands of other animals including mice, cows, goats, pigs, cats, sheep, horses, monkeys, and even endangered species have been cloned in laboratories around the world. As scientists gain increased understanding of the reprogramming process, cloning technology becomes increasingly efficient.

The cloning of embryos, by itself, serves no therapeutic purpose. But within a year of Dolly's public debut, James Thomson at the University of Wisconsin developed a complementary technique for growing unlimited quantities of the cells located at the core of a naturally conceived five-day-old human embryo.[19] These incredibly malleable cells are called *embryonic stem cells* or *ES cells*. Thomson mastered the technology required not only to keep these cells growing and dividing in a laboratory petri dish, but also to trap them in a Peter Pan-like embryonic state in which they retain their full potential. The power and future promise of ES cell technology lies in the ability of scientists to flip large quantities of these cells out of embryonic never-never land and into a specific tissue type or organ needed by a particular patient at a particular time. To accomplish this task, biomedical scientists must identify and learn how to mimic the molecular signals that are involved in the natural process of differentiation.

It is when embryo cloning and stem cell technologies are combined that the remarkable potential for an entirely new array of medical therapies is achieved. Traditional organ transplantation therapies have always been limited by both the scarcity of donors and the fact that a normal immune system is prone to see a donor organ as a foreign organism, which must be rejected and destroyed. Even so-called genetically well-matched donors and recipients are not perfect, which is why the failure rate is still quite high although all patients take immunosuppressive drugs. The only perfect match is between identical twins (who are usually not available), or between individuals and *their own tissue or organ*. A solution to both the scarcity problem and the rejection problem could be achieved through embryo cloning followed by stem cell derivation for therapeutic purposes, so-called *therapeutic cloning*. In the future, when a person suffers from a disease in a particular organ or tissue, one of her healthy cells would be used to create a cloned embryo for the production of ES cells, which would be converted into the required tissue. Since the replacement material would have the same genetic constitution as the patient (indeed, it really would be her own tissue or organ), the immune system would recognize it as part of its "self" and leave it untouched.

Experimental results have begun to yield hard evidence for the amazing promise and versatility of stem cells to regenerate a large array of different tissues and organs. What's truly stunning about the litany of advances is that none of it could have been imagined a decade earlier, and nearly all of it has occurred in a period of just five years. This is a lightning pace compared to the decades it used to take to develop new medical treatments, and it's just the beginning. With the combined use of stem cell and genetic modification technologies, clever scientists could someday attack every disease that harms human beings, but only if science, religion, and politics can find common ground.

Western Backlash and Asian Opportunity

Human embryo research and the genetic modification of plants are the two most contentious forms of modern biotechnology. And yet many people in Western countries are morally opposed to one while, in principle, they are accepting of the other. The opposition to human embryo research is rooted in the traditional Christian belief that a higher spiritual authority—God—creates each individual human being *in his image*. In this context, embryo research is immoral not simply because it may destroy a human life, but because the destruction of God's work violates God's plan.

The willingness of traditional Christian believers to accept the potential benefits of genetically modified crops also finds roots in the Bible. In particular, only man is said to be created in the image of God. All other living things are put on earth for man's benefit, and man is specifically given "dominion" over them. So while plants and animals may have been God's creations originally, God has since delegated responsibility for their upkeep to man. In this context, GM crops are not viewed as inherently good or bad. They can be evaluated, instead, based on a rational assessment of costs and benefits. Of the ten countries with the largest areas of GM crops under cultivation in the year 2004, six were in the western hemisphere.[20] Outside of North America (the United States and Canada), the four countries in this class—Argentina, Brazil, Paraguay, and Uruguay—are all dominated by powerful Catholic hierarchies. In addition, the Asian countries of China and India are ranked at numbers 5 and 7, respectively. But notably absent from the top ten GM producers is any country from Europe. The European rejection of GM crops, I argue, is a consequence of Europe's Christian roots combined with its current rejection of traditional Christian beliefs.

The results of a 1998 survey of religious beliefs in European countries indicates how far Europeans have moved away from their Christian traditions.[21] In Sweden, Switzerland, Norway, Germany, the Netherlands, France, and the United Kingdom, less than half the population holds a strong belief in a traditional Christian version of God. By comparison, a strong belief in God is professed by 78 percent of Americans and 91 percent of Chileans (the only South American population covered in this survey). Although the level of atheism and agnosticism in Europe is relatively high, another quarter or more of the population of many western European countries (Sweden, Switzerland, Austria, Norway, and western Germany) "don't believe in a personal God," but "do believe in a Higher Power of some kind."

It appears that many western Europeans are in desperate need of a substitute to fill the spiritual void left behind when the God of the Bible is rejected. But Western culture is permeated by Judeo-Christian monotheism, and so the substitution is made most easily through a transformation of traditional Christian beliefs into a *post-Christian* religiosity. The sacredness of the material human body—symbolized in Jesus—morphs into the sacredness of a material Mother Nature. God's master plan for humanity becomes Mother Nature's master plan for the whole-earth biosphere. Earth's creatures are now viewed as component parts of Mother Nature's body; if we engineer her genes—the modern analogue of a singular higher spirit—we are liable to upset the natural order.

An example of post-Christian European spirituality is etched into the Constitution of Switzerland through an amendment demanding respect for *l'intégrité des organismes vivants*, the integrity of living organisms, and *Würde der Kreatur*, the dignity of creatures (actually *living nature as a whole*). The amendment was not just a call to alleviate animal suffering or to prevent animal breeding for food (since most who voted for the referendum eat meat). Instead, a majority of the Swiss people felt that their picture-perfect valleys of well-tended meadows, neat farms, and grazing cows represented a *natural order* that had to be preserved at a deeper spiritual level. It doesn't matter that every component of this picture is a direct result of human intervention into a previous natural order that has long since disappeared.

The spiritual traditions of the East are diverse, but they all share certain Hindu-Buddhist-derived foundations in contradistinction from Western monotheism. There may be many gods or no gods (depending on the semantic distinction between gods and divine spirits), but there is no master creator in the East, nor is there any master plan that we can violate. If no master—or master plan—of the universe exists, the injunction to not "play God" has no meaning or suasion. Furthermore, in the Eastern worldview, bodies are discarded when they wear out, and their spiritual inhabitants reincarnate within new material beings. In this context, while biotechnology may affect the material, it can't touch the spirits, every one of which has existed from the beginning of time, and will go on existing forever no matter what we do.

The consequences of the divide between Eastern and Western spiritual frameworks may be dramatic for the future of biotechnology. Singapore and South Korea do not have the land to pursue GM crops, so they are focusing their energy on embryo research. Singapore has far more invested in an economic future devoted to stem cells and other forms of human biotechnology. Starting from scratch in the year 2000, the Singaporean government hatched a plan to invest and attract $3.5 billion into a massive biotech complex called Biopolis with two million square feet of space to house two thousand university, government, and industry researchers. Its scientific advisory committee reads like a Who's Who of the most prominent molecular biologists and biomedical scientists from the United States and Europe who are all Nobel Laureates, university presidents, or directors of major research institutes. And world-class Western scientists have been attracted to this hypermodern English-speaking city-state to head-up the major Biopolis research institutes including Edison Liu, the former head of the U.S. National Cancer Institute, and Alan Colman, a member of the scientific team that cloned Dolly the sheep. Johns Hopkins University Medical School, based in Baltimore, Maryland, is building a whole new Singaporean division devoted to research likely to be hampered by American religious sentiments.

India, on the other hand, will focus primarily on GM crops, while China is cleverly leveraging all biotech fields from embryo research, to GM crops, to human gene therapy technologies. They have made great efforts to bring home many expatriate scientists from the United States and Europe. The Chinese claimed to have an initial concern about GM crops, but most likely, that was a tactic to prevent American companies from entering their market. For economic and political reasons, they have focused on the genetic engineering of rice, knowing that this crop is not a European staple, and that other Asian countries are more spiritually accepting.

Generally, non-Christian Asians feel comfortable with all forms of biotechnology, and their governments are poised to leap ahead of Western countries in research and development of plant, animal, and embryo engineering. The economic ramification of cultural differences in spirituality may be enormous: a future in which the West dithers as Asia becomes dominant in both the science and commercialization of biological processes.

Notes

1. B. D. Smith, "Documenting Plant Domestication: The Consilience of Biological and Archaeological Approaches," *Proceedings of the National Academy of Sciences of*

the United States of America 98 (2001): 1324–1326; N. V. Fedoroff, "Agriculture Prehistoric GM Corn," *Science* 302 (2003): 1158–1159.

2. Consider B. M. Fagan, *World Pre-History: A Brief Introduction* (New York: Harper Collins College Publishers, 1996).

3. See Fedoroff, "Agriculture Prehistoric Corn,". 1158–1159.

4. See Smith, "Documenting Plant Domestication," 1324–1326.

5. See P. Savolainen, Y. P. Zhang, et al., "Genetic Evidence for an East Asian Origin of Domestic Dogs," *Science* 298 (2002): 1610–1613.

6. See B. Hare, M. Brown, et al., "The Domestication of Social Cognition in Dogs," *Science* 298 (2001): 1634–1636.

7. H. G. Parker, L. V. Kim, et al., "Genetic Structure of the Purebred Domestic Dog," *Science* 304 (2004): 1160–1164.

8. See A. Beja-Pereira, G. Luikart, et al., "Gene-Culture Co-Evolution between Cattle Milk Protein Genes and Human Lactase Genes," *Nature Genetics* 35 (2003): 311–313.

9. J. M. Kijas and L. Andersson, "A Phylogenetic Study of the Origin of the Domestic Pig Estimated from the Near-Complete mtDNA Genome," *Journal of Molecular Evolution* 53 (2001): 302–308.

10. See J. Diamond and P. Bellwood, "Farmers and Their Languages: The First Expansions," *Science* 300 (2003): 597–603.

11. Consider R. J. Braidwood, *The Near East and the Foundations of Civilization* (Eugene: Oregon State System of Higher Education, 1952).

12. See Paul Ehrlich, *The Population Bomb* (New York: Sierra Club-Ballantine, 1968).

13. G. S. Khush, "Green Revolution: The Way Forward," *Nature Reviews Genetics* 2 (2001): 815–822.

14. J. Ross, "The Organic Farmer's Story," *The Scotsman* (2001): 11.

15. See A. S. Van Laere, M. Nguyen, et al., "A Regulatory Mutation in IGF2 Causes a Major QTL Effect on Muscle Growth in the Pig," *Nature* 425 (2003): 832–836.

16. C. N. Karatzas, "Designer Milk from Transgenic Clones," *Nature Biotechnology* 21 (2003): 139–139; N. Smirnoff, "Vitamin C Booster," *Nature Biotechnology* 21 (2003): 134.

17. See Y. Kuroiwa, P. Kasinathan, et al., "Cloned Transchromosomic Calves Producing Human Immunoglobulin," *Nature Biotechnology* 20 (2002): 889–894; and B. Brophy, G. Smolenski, et al., "Cloned Transgenic Cattle Produce Milk with High Levels of Beta-Casein and Kappa-Casein," *Nature Biotechnology* 21 (2003): 157–162.

18. Ian Wilmut, A. Schnieke, et al., "Viable Offspring Derived from Fetal and Adult Mammalian Cells," *Nature* 385 (1997): 810–813.

19. See J. A. Thompson, J. Itskovitz, et al., "Embryonic Stem Cell Lines Derived from Human Blastocysts," *Science* 282 (1998): 1145–1147.

20. Consider C. James, "Preview: Global Status of Commercialized Biotech / GM Crops," ISAAA Brief No. 32 (Ithaca, N.Y.: International Service for the Acquisition of Agri-Biotech Applications, 2004).

21. See the International Social Survey Program, *Religion II* (Cologne: German Social Science Infrastructure Services, 1998).

FIVE

Jewish Philosophy, Human Dignity, and the New Genetics

HAVA TIROSH-SAMUELSON

Judaism and the Challenge of the New Genetics

We live in a revolutionary age that confronts us with unprecedented new possibilities, risks, and responsibilities.[1] The current confluence of advances in the life sciences (e.g., genomics, stem cell research, genetic enhancement, germline engineering), technology (e.g., robotics, nanotechnology, pattern recognition technologies), and neurosciences (e.g., neuropharmacology and artificial intelligence) signifies profound changes. While the new technologies have made important strides against devastating diseases, such as cancer, diabetes, and AIDS, alleviating human suffering and improving the human condition, they have also made it possible to go beyond healing and curing. Today the new genetics and its corollary technologies have brought us closer to the possibility of creating and modifying (i.e., clone and engineer) existing forms of life, including human life. Endowed with new knowledge and innovative technologies, human beings are gradually becoming their own makers, transforming their environment and themselves.

No one has been more acutely aware of the challenges of the new biotechnologies than Dr. Leon Kass, the outgoing chairman of the President's Council on Bioethics, whose essay commences this volume. With dogged determination, Kass has insisted we pay close attention to the social and cultural implications of the new biotechnology, engaging it philosophically and ethically, rather than blindly celebrating its promising potential. Kass avers:

> we now clearly recognize new uses for biotechnological power that soar beyond the traditional medical goals of healing disease and relieving suffering. Human nature itself lies on the operating table, ready for alteration, for eugenic and neuropsychic "enhancement," for wholesale redesign. In leading laboratories, academic and industrial, new creators are confidently amassing their powers, quietly honing their skills, while on the street their evangelists are zealously prophesying a posthuman future. For anyone who cares about preserving our humanity, the time has come to pay attention.[2]

Under Kass's leadership, the President's Council on Bioethics judiciously and thoughtfully evaluated a vast range of technologies, practices, and research agendas, including the Human Genome Project, stem cell research, germline engineering, cloning, and a host of assisted reproductive technologies for their social impact and implications for our self-understanding.[3] Concerned with the severing of the causal connection between sex and reproduction, which is changing the patterns of procreation and child-rearing, Kass has insisted that we address foundational questions: What kind of human beings do we wish to be? What kind of society do we wish to live in? What ideals, norms, and standards should guide us into the future?

Leon Kass is Jewish and his thinking about biotechnology is at least partially inspired by the Jewish tradition, especially his reading of the Bible.[4] However, on many of the currently disputed issues, such as in vitro fertilization, stem cell research, and cloning, Kass's cautionary stance toward biotechnology is not shared by the majority of Jewish jurists and medical ethicists. Unlike Kass, who argues for the "virtues of mortality," Reform, Conservative, and Orthodox rabbis tend to be strongly in favor of "more life, longer life, new life," as he correctly put it.[5] In contrast to Kass, Jewish legal authorities support assisted reproductive technologies, endorse stem cell research for medical purposes, and champion genetic screening, testing, and even genetic engineering.

Whereas believing Christians, especially Roman Catholics, feel deep anxiety about the current biotechnology revolution, the Jewish community has welcomed biotechnological advances and has taken an activist stance toward it.[6] Orthodox jurists evaluate each and every new technology not in terms of its impact on society at large, but in terms of its permissibility within the principles and reasoning procedures of Jewish law. Without a doubt cloning, including cloning to produce children, is the most radical manifestation of the new genetics, and yet Orthodox jurists are remarkably open to it. The dominant Orthodox view is captured by Rabbi David Bleich who states, "So long as the methods deployed in assisted procreation do not entail transgression of halakhic strictures, such methods are discretionary and permissible."[7] Strict halakhic reasoning has led Rabbi Azriel Rosenfeld to worry that cloning could destroy the family relationship, and the new methods could result in a person being born without a halakhic father. Rabbi Rosenfeld has concluded that cloning can be permitted because this reproductive method does not involve a sex act, and therefore, it is not halakhically forbidden. Another Orthodox jurist who has written extensively on biomedical issues, Rabbi Fred Rosner, initially did not approve cloning, since "cloning of men negates identifiable parenthood and would thus seem objectionable to Judaism," but in a later ruling he concluded that it is permissible.[8]

The positive attitude toward the new genetics and accompanying technologies, especially assisted reproductive technologies, is usually justified by appeal to the rabbinic portrayal of the human being as God's "partner in the work of creation." The idea is derived from Talmudic sources that teach that "three partners (God, man and woman) are required for the creation of a human being,"[9] meaning that humans cannot accomplish procreation alone and must receive divine involvement. Orthodox authorities reason that to be a "partner of God" means that humans have an obligation to improve and ameliorate what God has created because "God left it for human beings to complete the world."[10] Science and technology can and should be used for this purpose "as long as the act of perfecting the world does not violate halakhic prohibitions, or lead to results that would be halakhically prohibited." On the basis of this reasoning, Rabbi Abraham Steinberg asserts, "We are partners with [God] to improve the world. It is not an option—it is an obligation to continue to improve the world and do good for the world," which leads him to conclude, "we are therefore permitted to interfere in nature, nay, we are obligated to interfere, obligated to improve the world."[11]

With their Orthodox cohorts, Conservative rabbis share a positive stance toward the new biotechnology, derived from the commandment to improve the world, which gives more emphasis to the commandment to heal the sick and prevent or alleviate suffering. The leading Conservative jurist and bioethicist, Rabbi Elliot Dorff, has asserted that "Jews have the duty to try to prevent illness if at all possible and to cure it when they can, and that duty applies to diseases caused by genes as much as it does to disease engendered by bacterial viruses, or some other environmental factors."[12] On the controversial issue of stem cell research, for example, Rabbi Dorff has stated:

> The Jewish tradition would certainly not object to such research; it should actually push us to do as much as we can to learn about these lineages so that hopefully one day soon we can help people avoid cancer, or, failing that, cure it. This attitude follows from the fundamental Jewish approach to medicine, namely that human medical research and practice are not violations of God's prerogatives but, on the contrary, constitute some of the way in which we *fulfill our obligation to be God's partners in the ongoing act of creation.*[13]

On the even more problematic issue of cloning to produce children, Rabbi Dorff has cited various arguments against the technology on the ground that "human cloning threatens to undermine our humility and our sense of being limited in two important ways: it makes it possible to reproduce asexually, and it seems to promise immortality."[14] But after considering the various arguments against cloning he concludes "human cloning should be regulated, not banned."[15] Dorff allows cloning "only for medical research or therapy" and his view is derived from the requirement to help other people escape sickness, injury and death. Medical research serves the religious commandment to heal and to imitate God's healing power by extending cure to the sick.

The Jewish endorsement of reproductive technologies including research that will lead to cloning of humans is most notable in the state of Israel, where legal reasoning and public policies are openly informed by Jewish religious values no less than by secular considerations. As Barbara Prainsack and Ofer Firestine have noted, "technologies that are controversial in other parts of the western world, such as embryonic stem cell research, prenatal genetic testing and human cloning have not caused heated public debates in Israel and generally enjoy a liberal regulatory framework."[16] In Israel, biotechnology regulation is characterized by a relatively permissive approach and a low regulatory density. Moral objections to research that is highly controversial elsewhere in the world are virtually absent in the public debates in Israel. Because the Israeli government has viewed science and technology as matters of national priority,[17] scientists do not have to protect themselves from intervention by "nonscientists." As for human cloning, in 1998 the Israeli Knesset passed a law that bans human cloning and germline therapy for a period of five years, but that law still permitted research on the activation of cells and production of human embryonic tissue "without actually getting to a human clone." Dr. Abraham Steinberg, who heads the Medical Ethics Program at the Hebrew University and who helped develop the law against cloning, acknowledges, "even if the Knesset will ban cloning, if it will be invented in England or Japan, cloning will come anyway."[18] And several other halakhic authorities also ruled that cloning is probably halakhically permitted and may contribute positively to human health.

It is not difficult to explain why Jews today are quite enthusiastic about the new genetics and its accompanying biotechnology. Beyond the religious commandment to procreate and the obligation to heal the sick and alleviate or prevent suffering, the Jewish endorsement of the new genetics reflects a deep anxiety about the demographic weakness of the Jewish people today. The anxiety arises from a serious demographic crisis. The loss of one third of the Jewish people in the Holocaust combined with the fact that the Ashkenazi Jewry, the community that suffered most from the Nazi extermination policies, also exhibits a preponderance to inherited genetic diseases, such as, Tay-Sachs, cystic fibrosis, fragile X syndrome, Gaucher's disease, and breast cancer, deepens the Jewish resolve to remedy genetic ailments by resorting to the new genetics.[19] In post-industrialized, Western democracies the demographic threat to continued Jewish existence is further exacerbated by the combined effect of modernization, acculturation, assimilation, and social mobility, which have not only destabilized Jewish collective and personal identities but also contributed to the shrinking of the Jewish family. As a result of late marriage age, the choice to have few children, the common use of abortion among nonreligious Jews, and genetic and environmental factors that contribute to infertility, the Jewish family today is unable to replenish itself.[20] The current situation stands in marked contrast to the Jewish religious obligation to procreate, traced to God's command to the first humans to "be fruitful and multiply" (Gen. 1:28). In the state of Israel, moreover, these demographic pressures receive a special significance given the ongoing struggle between Israel and its Arab neighbors. According to the demographer Arnon Sofer, the non-Jewish population in Israel plus Gaza and the West Bank is expected to have outnumbered Israel's Jewish population by 2020 (8 million non-Jewish Palestinians in contrast to 6.6 Jews).[21] It is no wonder, therefore, that medical genetics is a recognized medical specialty in Israel where eleven clinical genetic centers offer genetic testing, genetic screening, and infertility treatment to a population of only six million, and Orthodox and Ultra-Orthodox rabbinic authorities provide religious justifications for a wide range of reproductive technologies.[22]

Since the threat to the continued existence of the Jewish people is real, utilizing technologies to increase birth rates among Jews is generally a good thing. However, it is also important to note that the new genetic technologies have also complicated the very question of Jewish embodied existence. What does it mean to be Jewish? Even if one agrees with the rabbinic norm that Jewishness is transmitted through the mother, Susan Martha Kahn has convincingly argued that "this transmission becomes less straightforward: is it the mother's egg that transmits Jewishness, or is it the act of gestation and parturition that makes a child Jewish?"[23] She explains:

> On this issue there is no consensus among rabbinic opinion: some rabbis argue that the act of gestation and parturition determines Jewishness, and those who wish to conceive a Jewish child may do so using eggs donated by non-Jewish women. Others argue the opposing view, that the egg transmits Jewishness, so in order to conceive a Jewish child, an infertile woman must receive an egg donated by a Jewish woman. By designating the egg as that which confers Jewishness, rabbis holding the latter opinion imbue genetic material with the powerful ability to create Jewish identity.[24]

Locating Jewishness in the body, as the new genetics compels us to do, is further problematized if we turn to consult the Jewish philosophical tradition.[25] Whereas the new genetics places the focus on the body, especially the genetic makeup of the body, the Jewish philosophical tradition identifies both humanness and Jewishness with the nonmaterial aspect of being human, namely, the soul, and even more specifically the rational aspect of the soul, the intellect. In our postmodern age, the intellectualism of Jewish philosophy has been criticized and denigrated as part of a wholesale critique of the so-called Enlightenment Project. I cannot explain or respond to this critique here, but only suggest that the Jewish philosophical tradition offers us a critical vantage point from which to consider the enthusiastic Jewish endorsement of the new genetics.

Contrary to the materialistic reductionism of the new genetics, the Jewish philosophical tradition regards the "image of God" (*tzelem elohim*) as that which makes us distinctly human and resists human tendencies to control, manipulate, and fabricate nature that God has created, including human nature. In the Jewish philosophical tradition, Jewishness is not a matter of biology, but of moral and intellectual perfection cultivated through the commitment to the Torah, broadly defined and philosophically interpreted. If the halakhic discourse resorts to the notion of partnership between humans and God to justify a pro-biotechnological stance, the Jewish philosophical tradition emphasizes the ontological gulf between the human and God and the responsibility of humans to protect the created order. For the philosophers, somatic existence is a necessary but not a sufficient condition for the attainment of our full humanity; the body is a means to an intellectual and spiritual end of human life, the knowledge of God to the extent God can be known by humans. In the Jewish philosophical tradition one can find further support to the cautionary stance of Leon Kass toward biotechnology.

This chapter discusses how four Jewish philosophers—Philo of Alexandria (15 BCE–ca. 50 CE), Moses Maimonides (1138–1204), Joseph Dov Soloveitchik (1903–1993), and Hans Jonas (1903–1993)—understood the meaning of creation in the "image of God" (*tzelem elohim*).[26] My choice of these philosophers requires some explanation since two of these philosophers—Philo and Jonas—are not considered authoritative within the Jewish legal tradition, and only one of them, Hans Jonas, spoke specifically about biotechnology; Philo and Maimonides lived long before biotechnology was a reality, and Soloveitchik reflected about science and technology in the context of his typological theology.

My selection of these four philosophers is deliberate, but my reasons for focusing on them are not arbitrary. First, these four thinkers illustrate the scope of the Jewish philosophical tradition, from antiquity to the present, covering the major locales of Jewish philosophical activities: Alexandria, in the first century, Spain and Egypt in the twelfth century, Germany in the early twentieth century, and America in the second half of twentieth century. Second, while it is true that Philo did not have authoritative status in the Jewish normative tradition, his philosophic exegesis of scriptures laid the foundation for the entire Jewish philosophical project. Many of the themes and the exegetical principles we identify with medieval Jewish philosophy, whose major exponent was Maimonides, were already stated by Philo.[27] Third, in terms of influence, at least three of these thinkers—Philo, Maimonides, and Jonas—exerted enormous influence on Western thought: Philo gave rise to the Christian philosophical exposition of scriptures, and eventually was viewed as one of the Church Fathers;[28] Maimonides exerted a deep influence on Thomas Aquinas and the reasoning of Christian Scholasticism; and Jonas is considered the intellectual inspiration of environmental philosophy and of the Green movement in Europe today.

Needless to say, the four Jewish philosophers do not exhaust the richness of Jewish philosophy, and Judaism also harbors diverse interpretations of the creation in the image of God, which could be cited in support of the new biotechnology.[29] Nonetheless, I want to highlight the Jewish philosophic tradition because so far Jewish reflections on biotechnology have been shaped by halakhists on strict legal grounds. Philosophy, by contrast, has been excluded from the conversation because contemporary Jews in general are either ignorant of it or consider it an imposition of "alien wisdom" on the authentic Torah-true tradition. This view is mistaken both historically and theologically. Jewish philosophy expresses the deepest religious posture of Judaism, the relentless commitment to truth, the passionate desire for knowledge, and the love of wisdom. Precisely because Jewish philosophy has been marginalized in our day it is incumbent upon us to recall the great epochs of Jewish philosophical creativity and learn from them. By seeking inspiration from the Jewish philosophical tradition, I will support further Kass's contention that what is at stake is not this or that medical procedure, not this or that technological invention, but rather a certain way of thinking about being human and the place of the human in the order of things. The four Jewish philosophers discussed here articulate deep insights about the meaning of being human, insights that the new biotechnology is poised to challenge.

Philo: Human Dignity and Uniqueness

Philo came from one of the wealthiest and most prominent Jewish families in Alexandria and belonged to the social elite that educated its members privately in institutions of its own.[30] Most Jewish members of the upper social class did not renounce Judaism and did not engage in public debates to denounce their lifestyle, but a few of them, such as Philo's brother, Alexander Lysimachus, did wish to assimilate into Hellenistic Roman society and consequently abandon Judaism. Philo's own son Alexander also became an apostate when his public career took him to the highest post of a Roman official in Egypt, that of a prefect. Philo was the acme of Jewish philosophical activity in Hellenistic Egypt, which had begun in the second century BCE, but spoke against Jews who interpreted scripture as philosophical allegory. At home in the culture of the Greek polis and thoroughly proficient in the philosophy of the major Greek schools of Plato, Aristotle, the Stoics, the Skeptics, and the Pythagoreans, Philo demonstrated how the Torah of Moses leads its adherents to the attainment of the ultimate end of human life—the "seeing" of God. Philo's exposition of scripture's truth was meant to prove the uniqueness and epistemological superiority of the Jewish tradition over other intellectual and religious traditions. Philo's philosophical interpretation of scripture presupposed that the truths of scripture and the truths of philosophy could not conflict with each other. His harmonization of Judaism and Greek philosophy was based on the original doctrine that the Torah of Moses is the *written copy of the laws of nature.*[31] Accordingly, the Torah of Moses contains the intelligible order of the world, which is accessible only to the intellect. The reader of scripture, therefore, cannot be satisfied with the mere literal sense of the text, but must seek out or expose the abstract, conceptual, or philosophical meaning of the text. It is only through the comprehension of that philosophical content of the revealed text that the human soul can perfect itself and attain the ultimate end of human life, the intellectual "seeing" of God.[32]

On the creation of the world and the creation of the human, Philo was mostly indebted to Plato, and Plato's dialogue, *Timaeus*, in which Plato depicts the creation of the world by a divine craftsman, the Demiurge.[33] According to Plato, the Demiurge wanted his creation to be as good as possible and therefore looked to an eternal and perfect model when he created the world. It is easy to detect the Platonic myth when Philo states, "God being God, knew beforehand that a beautiful copy would never be produced apart from a beautiful pattern, and that no object of perfection would be faultless which was not made in the likeness of an original discerned only by the mind."[34] Philo calls the perfect model that God had consulted Logos, a term that applies to the intelligible world of Ideas in its totality.[35] The Logos is the *first image of God* and it functions as the model (*paradeigma*) for the creation of the material universe. An intermediary between God and the physical world, the Logos is both an instrument (*organon*) that God used in creating the world as well as a cosmic power (*dynamis*) that is present in the world. The Logos binds all things together and causes them to cohere; it is the rational plan that governs the life of the universe. The world we experience through the senses is created as a copy of the intelligible paradigm, the Logos.

The Logos doctrine helps Philo explain the creation of humanity in the image of God. Genesis, as is well known, presents two creation narratives of the human: according to the first narrative (Gen. 1:1–2:3), the human was created in the image of God, and according to the second (Gen. 2:4–3:18), the human was created from the soil of the earth. For Philo, the first narrative pertains to the *Idea of humanity*, to the human mind and its ability to reason, which Philo considers the "heavenly human"; the second narrative depicts the "earthly human," the creature who is composed of a corporeal, mortal body and a divine, immortal soul. Whereas the "heavenly human" is a copy of God, or at least of the Logos, the physical, "earthly man" is made of loose material, which he, that is, Moses, calls "a lump of clay." Genesis 2:7 depicts the creation of a human being who was composed from a corporeal body and a "divine spirit."[36] As a Platonic thinker, Philo clearly privileges the divine spirit over the corporeal mind and holds that it is by virtue of its reason that the human "is allied to the divine Reason, having come into being as a copy or fragment or effulgence of that blessed nature, but in the structure of his body he is allied to the world."[37]

How can Philo's interpretation of creation in the divine image be relevant to our contemporary conundrum regarding biotechnology? First, Philo helps us to think about being human in a way that is between dualism of body and soul, on the one hand, and physicalist materialism, on the other. His subtle views are captured in the analysis of the word "image." In several places Philo employs the term "seal" (Greek: *sphragis*) as an equivalent of the "image" in the active sense of its force as a pattern. The Logos is called "the archetypal seal," "the seal of the universe," "the original seal" of which intelligible and incorporeal man is a "copy." The heavenly man too is called "an Idea, or type, or seal" because the earthly man is modeled after him. Moreover, Philo speaks of "a form which God has *stamped on* the soul as on the tested coin."

In Philo's Platonic reading of Genesis, then, God is portrayed as an artist, a craftsman, whose creative act involves impressing the abstract pattern, the Logos, unto the material substrate, as an artist would do in his workshop. The creative-artistic act of stamping a pattern unto a receptive material substrate happens most patently in the minting of coins or in impressing a seal. When Philo describes the creative act, he shifts between the motif of the "coin" and the motif of the "seal."[38] In the natural world impressing the divine image, or seal, unto the material substrate makes the divine pattern, or the Logos, be present in all levels of nature. In mammals, especially human ones, this impression takes place in the womb, which Philo calls "nature's workshop." If so, the world is inherently intelligible and its intrinsic structure is accessible to the human mind that is stamped in its image. Such a posture encourages humans to explore the structure of the world created by God as much as possible, since it is this exploration that makes us unique. All scientific exploration into the manifestation and structure of the created world are thus religiously sanctioned.

My attempt to draw a Jewish lesson from the writings of Philo may be rejected by some Jews on the grounds that Philo's philosophy and exegesis conflict with those of the rabbis, the framers of normative Judaism.[39] The precise relationship between Philo and the rabbis has been subject to intense scholarly debate, but we should note that Philo's association between the motifs of the seal and the coin is found in rabbinic sources that similarly wrestle with the meaning of creation in the divine image. For example, in Mishnah Sanhedrin

4:5 we read the following in the context of a legal discourse concerning testimony in capital cases: "For a person mints many coins with a single seal, and they all resemble one another. But the King of Kings of kings, the Holy One, blessed be He, minted all human beings with that seal of his with which he made the first person, yet not one of them is like anyone else. Therefore everyone is obligated to maintain, 'On my account the world was created.'"

Like Philo, the rabbis established human uniqueness and intrinsic worth by using an analogy: whereas earthly kings mint coins in which there is no difference between one coin and another, only God, the king of all kings, created human beings with the seal of the first human, but each one is distinct and unique, unlike any other human being. Although it is true that the rabbis did not share Philo's distinction between the "heavenly man" and the "earthly man" and did not elaborate a Logos doctrine, they did share the notion that God consulted the primordial Torah when He created the world, and their reflections about the role of Wisdom in the creation of the world parallel Philo's doctrine. For the rabbis, no less than for Philo the created order of the universe exhibits a rational order that is accessible to the human mind because the mind is created in "the image of God." Similarly, the rabbis like Philo associated the "seal" with Adam as in the phrase "the seal of the first Adam" (*hotamo she adam ha-rishon*), and the rabbis also depicted God as a creative artist, specifically a painter, who created the world by looking at a pattern or a form.[40]

The principle of human dignity and intrinsic worth is articulated in many rabbinic sources and one well-known text is Mishnah Avot 3:14. There we read:

> He [R. Akiba] would say, "Beloved is man for he was created in the image [of God]; still greater was the love in that it was made known to him that he was created in the image of God," as it was written, "*For in the image of God made he man*" [Gen. 9:6]. Beloved are Israel for they were called children of God; still greater was the love in that it was made known to them that they were called children of God, as it is written, "You *are the children of the Lord your God*" (Deut. 14:1). Beloved are Israel, for to them was given the precious instrument; still greater was the love, in that it was made known to them that to them was given the precious instrument by which the world was created, as it is written, "*For I give you a good doctrine. Do not forsake my Torah*" (Proverbs 4:2).

This pericope teaches not only that human beings are a unique species because they are created in the image of God, but *that human uniqueness lies in the fact that they possess a self-reflective consciousness of their own uniqueness.* We humans are a class of creatures endowed with reason whereby we exercise intellectual knowledge of the world and engage it accordingly. As self-conscious embodiment, we humans are not owned by our innate capacities because one of our capacities is an endowment where we can own ourselves. As self-reflective consciousness, we do not play out some fixed "script," let us call it the "script of the genes." Rather, we become actors and agents because part of the scripting and constitution of our personhood is our capacity to own our roles and create much of the "script."[41] For Jews the "script" refers to the relationship with God through the interpretation of Torah, the paradigm that God consulted when he created the world, very similar to the way Philo understood the role of the Logos. The rabbinic understanding of creation in the image of God is not far apart from that of Philo, notwithstanding recent attempts to liberate the rabbis from the burden of medieval and modern Jewish philosophical interpretations of their legacy.[42]

The fact that humans, and especially Jews according to Philo, have access to the intelligible structure of the universe does not give humans, including Jews, a license to do with the world what they see fit. While Philo considers the human to be superior to all other animals and asserts human mastery of the natural world, he also claims that humans have a *responsibility* toward the natural world and that they must act as *stewards of nature*.[43] Entrusted to take care of God's creation, the steward has to ensure the perpetuation of the natural world as created and not to interfere in nature simply to satisfy human whims. Philo was an astute observer of agricultural and farming practices of his day, and was aware of various techniques to improve crop yields through grafting or increase the herd through selective breeding. But Philo also understood that human interference with natural processes must be regulated by the Torah itself since it defines the limits for such interference. For Philo, then, the laws on mixed breeding and cross-fertilization illustrate why only the Torah is the written copy of the law of nature since the Torah protects and preserves the created order.

By reading Philo and the rabbis together, albeit selectively, I wish to highlight the similarities between their projects, since both understood that the biblical text could *not* be taken literally. The literal meaning of the word "image" (*tzelem*) was derived from the Old Akkadian and Old Babylonian *tzalmu*, which means "a statue, a bodily shape, a figurine, or relief drawing." In Near Eastern cultures, the term referred to the image of the king, which was placed in a captured city to represent the king and his law. Historically, when the biblical text asserted that the human being was created in the image of God it most likely meant that the human represented the divine presence in the world. On earth, humankind functions as a kind of viceroy of the mighty King of heaven and earth, ruling over all the animals. In seeking to fathom the meaning of the biblical text, both Philo and the rabbis went beyond the literal meaning of *tzelem* to identify the image not with governorship and politics but with understanding. For Philo, the rationality of the world matches the rationality of the human mind, and for the rabbis the focus is on possessing reflexive consciousness.

Rabbinic sources cite creation in the image of God to derive various moral lessons such as the gravity of hurting, offending, or killing a fellow human being who was created in the image of God and as the justification for the commandment to be compassionate and merciful to all creatures. These moral teachings can be drawn whether one interprets the image anthropomorphically (as the rabbis did) or whether one interprets the image non-anthropomorphically (as Philo did). The following passage from Yebamot 8.7 shows how the rabbis linked the image both to moral imperative and to procreation:

> R. Akiva says: "Whoever spills blood, such a one annuls the [divine] likeness, since it says: 'he who shed the blood of man his blood will be shed'" (Gen. 9:6). Ben Azzay says: "Whoever does not engage in reproductive sexual relations such a one sheds blood and annuls the [divine] likeness, since it says: 'for in the image of God he made man,' and it says 'and you be fruitful and multiply.'"

I am aware that this text can be cited in support of assisted reproductive technologies: if all of humanity constitutes the great divine body so that harming humanity is harming the body of God and not engaging in reproduction diminishes the dimension of the body of God, one might conclude that any technology that helps propagate the body of God should be permitted. If the body of Adam was the great plan for the unfolding of humanity, if the body itself forms a plan that unfolds through the furtherance of life, one could argue that humans should be actively involved in such an endeavor. But the same text can also be interpreted negatively: if the body of the primordial Adam is a manifestation of God, and if we are a diminished form of the primordial creation, who are we to interfere in the beauty and glamour of the primordial creation? It seems to me that the Jewish philosophical tradition at least includes a voice that speaks against interference in the created world in order to conserve the order created by God.

Maimonides: Human Dignity and Intellectual Perfection

Philo's writings in Greek were preserved by the church and did not become part of normative Judaism, but many of the themes and exegetical moves made by Philo in the first century were elaborated by Moses Maimonides in the twelfth century, although Maimonides did not have direct access to Philo's works. The similarity between their positions arises from the nature of the project in which they were engaged and the commitment to the basic insight that to be Jewish means to seek the knowledge of God, to the extent God can be known by humans. Like Philo, Maimonides inherited a blend of philosophical traditions, including Platonic, Neoplatonic, Stoic, and Aristotelian, although it was the latter that dominated Maimonides' worldview. And like Philo, Maimonides held that the inner meaning of the Torah does not conflict with the truth of philosophy. Perpetuating the philosophic exegesis of the Bible initiated by Philo, Maimonides made clear how creation in the image of God pertains primarily to the intellect and how Jewishness lies in the cultivation of moral and intellectual virtues, culminating in the knowledge of God, to the extent that God is knowable.

Maimonides was undoubtedly the most important Jewish philosopher, jurist, and physician in the Middle Ages. Born in Cordoba, but living his adult life in Egypt, Maimonides was the leader of the Jewish community in Egypt, while functioning as the physician of the court of Al-Afdal, the vizier of Salah al-Din. His immense philosophic, scientific, and medical knowledge was culled from numerous literary sources in Arabic and Judeo-Arabic and mastery of the entire Jewish legal tradition culminating in his Code of Jewish Law, the *Mishneh Torah*.[44] Maimonides' *Guide for the Perplexed* is not a commentary on the Bible but a guide to the principles that should inform Jewish understanding of scripture as a philosophic text. The first part of the *Guide* constitutes a philosophical dictionary of sorts, in which Maimonides tells his readers how to interpret specific biblical words that are inherently ambiguous.

In the *Guide* I:2 Maimonides explicates the meaning of the terms "image" (*tzelem*) and "likeness" (*demut*) in relationship to God. Maimonides adamantly rejects the notion that the image of God refers to the physical body. Such a reading is considered by him to be not only naïve and philosophically uninformed but also religiously dangerous. By thinking that the image refers to the body of God, we attribute embodiment to God, which Maimonides emphatically rejects. God is not a body, nor a force in a body; God does not have a body, and no form of embodiment can be ascribed to God.[45] Such radical and consistent rejection of embodiment was meant as a critique of Jewish mystical speculations about the body of God cultivated among rabbinic circles in Baghdad in which the texts that constitute the *Shiur Qomah* tradition were edited.[46] Maimonides emphatically rejected these speculations on the ground that they compromise the radical otherness of God and lead to idolatry. However, the tradition flourished in medieval kabbalah.

The radical otherness of God, which Maimonides teaches in his negative theology, makes impossible the notion that humans and God are partners. For Maimonides, God is unlike anything else, a Necessary Being whose existence is identical with His essence. God's radical otherness means that He cannot be defined or grasped by human categories. All positive attribution to God is inadequate; God is the wholly transcendent God, not dependent upon any other existence and not even related to any other existence (*Guide* I:52). We cannot say who or what God is but only what God is not, that is to say, we can approximate a proper understanding of God only if we interpret any term we apply to God negatively. God is completely different, the otherness of his existence can only be known by what He is not or by what he does. Maimonides' radical negative theology is emotionally difficult to accept, because it is too cerebral and emotionally austere. Even he violated the radical otherness of God in one primary respect: in regard to the act of thinking, it is possible to speak positively about God. God is an intellect, a mind, engaged in eternal self-contemplation (*Guide* I:68). God eternally contemplates Himself, that is to say, contemplates eternal, necessary truths.

It is this aspect of God that constitutes the meaning of creation in the divine image according to Maimonides. The "image" has nothing to do with the physical shape of humans and/or God but with the act of thinking. Maimonides interprets the biblical creation narrative as an elaborate philosophical parable about the human intellectual potential. The word "*adam*" is both a proper name of one individual within the human species as well as a general term for the human species. In the latter sense, the word "*adam*" stands for the species as a whole as well as the *form* of the human species whose attainment constitutes the ultimate end of human life. The word "form" is a technical philosophical term that Maimonides understood in accord with Aristotle's metaphysics of matter and form. The form is that which makes a thing what it is; matter is that which determines that a thing is. In Maimonides' Aristotelian metaphysics, matter and form always exist together, but form is ontologically superior to matter.

How does the metaphysics of matter and form enable Maimonides to understand the meaning of the image of God? According to Maimonides, the "image" refers to the form of the human species, which is the intellect. When fully actualized the form of the human species is not connected to a body and does not depend on a body. The human form "knows and apprehends the Intelligences that exist without material substance; it knows the Creator of all things and it endures forever." In Adam prior to the sin of disobedience this form was fully realized, so that Adam engaged in contemplation of truth and was able to make distinctions between truth and falsehood. However, the sin of disobedience meant a shift in orientation: from contemplation of truth, Adam shifts to making moral distinction between good and bad.[47]

Why did this shift come about and what does the change in epistemic level signify? Maimonides would give a gendered answer to this question whose details cannot be explained in this context.[48] Briefly put, in the Garden of Eden there was only one being, Adam, comprised of matter and form; the latter is "male" and figuratively called "Adam" and the former is female and figuratively called "Eve." The Torah tells about "Adam" and "Eve" to teach the interdependence of matter and form. The sin of disobedience was following the temptations of the body, which is matter, as opposed to engaging in intellectual activity which is the formal aspect of the human. The result of the sin was a diminution of the human stature not physically but intellectually and morally. The human being as we know it is *not* identical with the primordial Adam, but it is the primordial Adam that sets the standards or ideals that humanity must aspire to regain. The Torah teaches that humans are created in the "image of God" in order to instruct us that we are created with the capacity to comprehend abstract necessary truths and to contemplate them. To be a human being, then, is not simply a matter of belonging to the human species; it is not about biological existence alone. Rather, it means *to possess the form of the human species*, namely, attain the intellectual perfection that consists in the activity of contemplating eternal truths. It is important to note that according to Maimonides *not every specimen of the human species is, truly speaking, human*; some are more human that others because they engage in the activity that is the perfection, or excellence of humanity.

Already the rabbis identified "creation in the image of God" with the command to imitate God and to follow His ways.[49] But how can one imitate God according to Maimonides, especially if God is so radically different? Maimonides provides two complimentary answers to this question that correspond to his understanding of the human as a composition of matter and form. First, although we cannot know God's essence, we can see the manifestation of God's essence in the created order of the universe. The way God manages the universe tells us about the attributes of divine actions. "To walk in God's ways," means to imitate divine actions as manifested in the natural world. The first task for all human beings is then to study the natural world (namely, to become proficient in the study of physics) and then to express the correct understanding in our moral and political actions. This is how Maimonides interprets the prophecy of Moses, the human being who reached the highest degree of understanding of God possible for humans and who expressed his knowledge in the form of law. The inner meaning of the Torah of Moses thus matches the structure of the universe, very much as Philo already believed.[50]

Jews who live by the Torah of Moses, Maimonides asserts, can accomplish the difficult task awaiting humans, namely, attaining moral and intellectual perfection. To resemble God means first to perfect one's character traits through correct ethical activity as taught by both Aristotle and the rabbis. Having achieved moral perfection by acting correctly in the social sphere, one then proceeds to acquire intellectual perfection culminating in the attainment of the state that the rabbis designated as "the world to come." This is a state of cognitive perfection in which the human rational capacity is fully realized and the human intellect is engaged in the most blissful activity of all: the intellection of truth. For Maimonides, this blissful act secures immortality, since the cognitive content of the human mind is itself eternal.

Maimonides' arguments in support of his cognitive understanding of human perfection requires further discussion of his theory of knowledge, which goes beyond the scope of this chapter.[51] Suffice it to say that for three centuries the theory of knowledge was wholeheartedly accepted by all Jewish philosophers, although it was heartily debated and rejected by Jewish intellectuals who endorsed kabbalah. By the fifteenth century, Hasdai Crescas disproved the theory by pulling apart its Aristotelian foundations, even though Crescas was unable to fully disengage himself from it. We are concerned not with the details of Maimonides' epistemology but with the vision he posited for us. According to that vision, to be created in the image of God means to engage in the ongoing pursuit of perfection by acquiring the moral and intellectual virtues. Through this pursuit humans can in principle overcome their mortality and at least some of them can enjoy the bliss of immortal life that consists of contemplation of truths.

Maimonides' intellectualism does not undermine the importance of the body in his anthropology. After all, Maimonides was the most celebrated physician of his day, who attended to the medical needs of the Muslim court, as well as to the needs of his fellow Jews in Cairo. The proper maintenance of the human body was a proper concern of Maimonides' extensive medical knowledge, which was put to daily practice. Maimonides was keenly aware of the interdependence of physical and mental aspects of human beings; if his medical practice concerns the health of the body, his philosophical knowledge pertains to the health of the soul; the two were intrinsically linked. However, Maimonides philosophical endeavor makes clear that he privileged the human intellect and that he considered the perfection of the intellect through the study of philosophy to be the ultimate end of human life. While proper control of the sexual impulse was viewed by him to be indispensable to human health, Maimonides considered the proper functioning, or health, of the body to be *a means to an end and not the end of human life*. Even the acquisition of moral virtues through action in the mean does not constitute the ultimate end. Survival and perpetuation of the human species is not what life is about; rather, to fully gain the Form of the human species one must engage in intellectual activity.

So how does Maimonides' legacy help us to reason about the current challenges poised by the new genetics? First, Maimonides makes it clear that to be human cannot be reduced to our mere biological existence, even though the biological or material dimension of human existence is a necessary condition to human existence. By identifying the intellect as the mark of humanity, Maimonides provides a very important vantage point from which to evaluate the danger and promise of the new genetics. Reproducing bodies is not what makes humans human; it is the actualization of the human rational potential that defines our humanity. Hence, we should evaluate our culture and our institutions in light of the task given to us by God, as Maimonides interpreted that task. Maimonides' privileging of the intellect does not fall into the trap of dualism. Rather, following Aristotle, Maimonides (and the Jewish philosophers who followed him in the late Middle Ages) understood the intrinsic connection between our mental functions and our biology. The interdependence of the cognitive and noncognitive aspects of humans are clearly illustrated in the process of moral conditioning, which is the beginning of our pursuit of human perfection. Maimonides' intellectualism, then, stands in sharp contrast with the current preoccupation or even fixation with Jewish biological survival. By the same token, Maimonides' intellectualism conflicts with reductionist materialism, characteristic of modern philosophy, science, and technology. I take Maimonides' intellectualist vision of humanity to function as an ethical guide that tells us to put the pursuit of truth at the top of our concerns, subordinating all other human pursuits to it. Such commitment, Maimonides amply argues, does not stand in conflict with Jewish religious faith; in fact, it is the highest expression of that faith properly understood.

Much could be gained today by recourse to the Maimonidean-Aristotelian understanding of moral perfection through the acquisition of virtues, a discourse that has fallen into oblivion in modern and contemporary Judaism to the detriment of us all.[52] In regard to one moral virtue—humility—Maimonides followed the rabbis rather than Aristotle.[53] But it is precisely this virtue that the new genetics and biotechnology more generally seem to ignore. Maimonides reminds us that nothing created can become God, let alone play God's exclusive and unique role as Creator. There is a close nexus between Maimonides' emphasis on humility and his insistence that we cannot know God or be God. This reminder may help us move more cautiously and slowly, knowing that we do not know what God knows and that we cannot reverse or correct any mistakes we make. Finally, Maimonides suggests that the human desire for transcendence is in principle attainable not through the physical body or material embodiment but through intellectual perfection. To be like God is ultimately to think, to contemplate, and to reflect about eternal truths. Would Maimonides be an avid promoter of Artificial Intelligence? I venture to say probably not, since Artificial Intelligence is not human but mechanical, and Maimonides was deeply committed to protecting the boundaries of the human species. In this regard he was guilty of what environmentalists today deride as the sin of "speciesism" and probably he would be upset with the current desire to blur the qualitative difference between humans and animals or between humans and thinking machines. Maimonides wanted to protect human distinctiveness from which follow moral obligations and political guidance.

Maimonides' intellectualist vision shaped Jewish philosophy for the subsequent centuries, as much as it also gave rise to rigorous debate about the meaning of being human and the desired degree of interaction between Judaism and surrounding cultures.[54] During the early modern period (sixteenth to eighteenth centuries), the Maimonidean worldview was eclipsed by kabbalah, although during the late Middle Ages the two intellectual programs were closely intertwined and cross-fertilized each other.[55] For the kabbalists, the "image" does not stand for the human intellect, that is, the form of the human species, but an astral body that enables the human to communicate with supernal powers and thereby to receive supra-cognitive knowledge.[56] More broadly, kabbalistic anthropology understands the psychophysical complex we call "human being" to mirror God, a unity within a plurality of ten dynamic forces known as Sefirot.[57] In the kabbalistic worldview the earthly human body serves as a vehicle to the interaction with God through the observance of the commandments that link each limb of the human body to an aspect or attribute of the Sefirot. While the kabbalists endowed the human body with sacramental significance, it is important to remember that the physical world and corporeality in general is but a reflection of the divine essence.[58] The kabbalists, therefore, did not glorify or worship the physical body, but viewed it as a medium for the spiritualization of the human through the observance of the commandments. As a program for self-spiritualization, kabbalah removed the human being from the natural world even when it interpreted the symbolic meaning of biological functions and physical action. Since nature was understood to be ultimately a linguistic text comprised of the units of divine energy (namely, the letters of the Hebrew alphabet), only the kabbalist was believed to possess the privileged knowledge of the linguistic code of creation. As a result of this worldview, kabbalah generated two different attitudes toward the natural world: on the one hand, kabbalah led to a "hands-on" approach to the natural world according to which mastery of the linguistic code of nature enables the kabbalist to create new entities, namely, engage in magic. This approach is exemplified in the famous legend about the creation of the artificial humanoid, the golem,[59] ascribed to the kabbalist R. Judah Loew of Prague (Maharal) (d. 1609), who was deeply immersed in the study of natural philosophy and the occult arts as were other Jewish and non-Jewish scholars associated with the court of Rudolph II in Prague.[60] On the other hand, kabbalah diminished interest in the physical world, since kabbalists focused their creative energies on the interpretations of the symbolic text. Be that as it may, both kabbalah and rationalist philosophy were seriously challenged by the new mechanistic worldview ushered in by the Scientific Revolution of the seventeenth century.

Joseph Dov. Soloveitchik: Human Dignity—Majesty versus Humility

Contrary to the Aristotelian assumption of all Jewish philosophers, the mechanistic worldview, articulated by Francis Bacon and René Descartes, denied that the physical world possesses an inherent *telos* and undermined the kabbalistic notion that the physical world is a symbolic text that reflects divine reality. For the modern philosophers the physical world of nature is indeed governed by set laws, the laws of nature, which God had established but in whose management God does not intervene. The modern worldview problematized Jewish existence because it challenged the very foundations of Judaism, namely, the belief in the divine Creator who brought the world into existence, who governs the world with wisdom and care, and who has entered a special relationship with Israel, the Chosen People, to whom God gave the Torah. Political changes, including the rise of the nation-state, liberalism, and democracy further exacerbated the challenges of modernity for Jews, since they undermined Jewish particularistic existence and led to the dissolution of Jewish communal structures. Jews responded to the crisis of modernity in various ways: those who favored integration into modern society, demanding equal, civil rights, endorsed the modern outlook as articulated in the Enlightenment and some of them were prepared to abandon traditional Judaism in order to assimilate. Most Jews, however, were not ready to forsake the ancestral tradition and sought rapprochement with modernity by offering various programs for the modernization of Judaism. Some reformers doubted the authority of the rabbinic tradition and called for wide-ranging ritual changes, while others defended the authority of halakhah while affirming the desirability of entering the modern world. Founded by Raphael Samson Hirsch in Hamburg, Germany, during the 1840s, this strand of modern Judaism known as Neo-Orthodoxy, later referred to as Modern Orthodoxy, would employ Kant's deontological ethics and his ideal of the noumenal as a point of departure for a modern philosophy of Jewish law.[61] In the United States during the twentieth century, Modern Orthodoxy was led by Rabbi Joseph B. Soloveitchik. Although he did not speak directly about biotechnology, his religious philosophy contains deep insights concerning modern science and technology.

A scion to an old dynasty of rabbis that led Eastern European Jews, Soloveitchik was tutored privately by his father in the rigorous dialectic of halakhic logic associated with the Yeshivah of Brisk, while taking private lessons that enabled him to attain a diploma in the liberal arts from the gymnasium in Dubno. In 1924 he entered the Free Polish University in Warsaw, where he studied political science. In 1926 Soloveitchik moved to Berlin to study in the Friedrich Wilhelm University, in the German Institute for Studies by Foreigners. In Berlin he studied philosophy and economics and was deeply attracted to the Neo-Kantian Marburg School led by Paul Natorp and Hermann Cohen, writing his doctoral dissertation on Hermann Cohen, because he found his mathematico-scientific idealism to be conducive to a philosophy of halakhah.[62] Invited to become the Chief Rabbi of Boston, Soloveitchik left Germany in 1932 and settled in Boston, from whence he shaped the intellectual program of Modern Orthodoxy in the United States. In 1937 he founded the first Hebrew Day School for boys and girls, which, not coincidentally, was named the Maimonides School, although the precise relationship between Soloveitchik's thought and Maimonides remains a disputed matter among his interpreters.[63] For more than half a century

Soloveitchik trained, ordained, and inspired scores of Orthodox rabbis through his senior Talmud seminar in the Isaac Elchanan Theological Seminary, which was affiliated with Yeshivah University, and the course on Jewish philosophy at the university's Bernard Revel Graduate School.

Like Maimonides, Soloveitchik wrote in Hebrew as well as in the vernacular, but the linguistic choice of a given text does not neatly correspond to a particular audience. For example, Soloveitchik's essay "Halakhic Man" was published originally in Hebrew in the journal *Talpiyot*, whose readership was not limited to observant Jews, whereas the famous essay "The Lonely Man of Faith" was published in English in the journal *Tradition* whose readership is generally Orthodox Jews.[64] That essay, moreover, originated in a series of lectures delivered in 1957–1958 for the National Institute of Mental Health, addressing Jews and non-Jews. Although Soloveitchik was the spiritual leader of a particular strand of Judaism, and was quite critical of its Reform and Conservative variants, his reflections were not limited to denominational ideology; rather, they amounted to philosophy of religious personality as well as philosophy of Jewish law. As an Orthodox Jew he maintained that normative Judaism offers the best way to experience the inevitable conflict of the human condition, exacerbated by Modernity but not occasioned by it.

The opening sentence of Soloveitchik's essay "Majesty and Humility" succinctly captures the human condition: "Man is a dialectical being; an inner schism runs through his personality at every level."[65] The conflict does not reflect a new historical situation between modern secularism and a faith commitment, nor should it be read as a summary of Soloveitchik biography;[66] rather, the conflict is rooted in the very way God created human beings. Created in the divine image, the human personality mirrors two conflicting divine attributes: majesty and humility. Both attributes are sources of human dignity, but in opposing ways: whereas the majesty of God leads humans to "rule, to be king, to be victorious,"[67] resulting in scientific knowledge, the dominion of nature, and the curing of diseases, humility is the divine attribute of "withdrawal and retreat" that facilitates love relations and the creation of faith communities.

For Soloveitchik science and technology are rooted in the human imitation of God but they express only one dimension of human uniqueness and dignity. In the modern era, human beings mistakenly reduce dignity solely to that realm of human activity, forgetting the other divine attribute, humility; God is not only a victorious ruler, but also the Creator who was able to impose a limit on Himself in order to make room for nondivine reality. The origins of this doctrine go back to a rabbinic source (Exodus Rabbah 34), but the notion of divine "contraction" or "withdrawal" (*tzimtzum*) is associated mainly with the kabbalah of Isaac Luria (d. 1572) who taught that the creation of the world became possible because the infinite God (*Ein Sof*) withdrew into Himself "to make room for a finite world."[68] Although Soloveitchik does not elaborate on the kabbalistic doctrine, he cites it in support of the notion that God's self-withdrawal "requires man to withdraw." On the basis of the kabbalistic doctrine, Soloveitchik assesses the modern human condition as follows: "Modern man is frustrated and perplexed because he cannot take defeat. He is simply incapable of retreating humbly. Modern man boasts quite often that he has never lost a war. He forgets that defeat is built into the very structure of victory, that there is, in fact, no total victory; man is finite, so is his victory. Whatever is finite is imperfect; so is man's triumph."[69] The problem of the modern person is not the existence of a tension between majesty and humility, but the refusal to accept that both constitute the meaning of being human. It is this denial that makes it impossible for the modern person to "retreat humbly" and enter into personal relationships with God and with other humans that constitute a faith community.

In his famous essay "The Lonely Man of Faith," Soloveitchik presents these insights as a typology of two foundational orientations, which he names "Adam the first" and "Adam the second." The reference, of course, is to the two creation narratives in Genesis 1–3. In the first narrative, God (*Elohim*) created the human being after "his own image" at the end of the six days of creation as a unity of male and female; according to the second narrative, God (*YHVH*) made the human out of the dust of the earth, breathed in him a "breath of life" (*nishmah hayyim*) and created a female counterpart. Like Maimonides, Soloveitchik takes the two biblical narratives to be complimentary aspects of one story, thereby ignoring modern biblical criticism that views them as two distinctive and even conflicting literary traditions. But whereas Maimonides read the biblical text in accord with Aristotelian metaphysics of matter and form, Soloveitchik interprets the biblical narratives as a religious existentialist: the two narratives are complimentary aspects of the inherently dialectical human condition, two different ways that human beings can understand themselves and their relationship to the world and to God.[70]

The first orientation, "Adam the first," "sees the world as an object to be mastered" and derives dignity from science and technology that make possible his overcoming of his vulnerability to nature. Desiring to overcome helplessness, human beings interact with each other on a utilitarian, functional basis mastering the external world, overcoming poverty, hunger, and other natural limitations. The technological "Adam the first" senses his humanity and dignity through the unleashing of energy and power, but his posture has little to do with awareness of God; it reflects instead the awareness of the human as being part of the cosmos. Appropriately the first creation narrative refers to God as *Elohim*, the name that captures God's power and governance of nature.[71] By contrast, the orientation of "Adam the second" leads human beings to experience the world not as a reality to be controlled, "but as a reality to be experienced with a sense of wonderment, puzzlement, and surprise."[72] From this perspective, human dignity is derived not from mastery and control of nature but from "the quest for purpose, meaning, and relationship" that the human finds in the encounter with another person, a Thou, through friendship, passion, and love. The second creation narrative, therefore, refers to God not in the impersonal name Elohim but in the proper name of the Tetragrammaton, signifying a personal relationship. The dignity of "Adam the second" is thus derived from personal relationship with God and with other humans, beginning with the relationship with the woman.

Entering personal relationship, however, requires self-limitation, imitating God's humility and self-imposed withdrawal. Although both orientations of "Adam the first" and "Adam the second" are divinely sanctioned, Soloveitchik notes, "Contemporary Adam the first, extremely successful in his cosmic-majestic enterprise, refuses to pay earnest heed to the duality in man and tries to deny the undeniable, that another Adam exists beside or, rather, in him. By rejecting Adam the second, contemporary man, *eo ipso*, dismisses the covenantal faith community as something superfluous and obsolete."[73] What Soloveitchik laments here is not the rise of secularism per se but the loss of human ability to truly imitate *humilitas Dei*. In modernity the victory of science and technology has brought about the loss of the "metaphysical polarity implanted in man as a member of both the majestic and covenantal community." Without the ability to imitate God's humility, modern humans are unable to make room for meaningful interpersonal relations, as much as they are not able to "to withdraw from rationalistic position" in order to freely assume the burden of divine laws. As David Hartman explains, according to Soloveitchik, "it is God who wants the man of faith to oscillate between the faith community and the community of majesty, between being confronted by God in the cosmos and the intimate immediate apprehension of God through the covenant,"[74] but modern humans have undermined that creative dialectic, resulting in spiritual impoverishment.

Soloveitchik elaborated his philosophical anthropology in a sustained commentary on Genesis 1–3, published posthumously in English under the title *The Emergence of Ethical Man*.[75] His exegesis is a masterly blend of the Jewish exegetical tradition and modern philosophy, especially of Kant, Bergson, and Cohen. Along with traditional Jewish sources, Soloveitchik considers the human being the "apex of the bio-pyramid,"[76] the "most developed form of life on the continuum of plant-animal-man" and as such has close kinship with nature. Reiterating the biblical outlook, he avers, "As long as man lives within the bounds set by his Creator, which accentuate his naturalness, he remains *ben adam*, the son of Mother Earth, and may claim asylum in her lap."[77] The human being is no mere biological reality, but also a being who is able to make ethical choices and receive divine commands, which render the human radically different from other natural entities and justifies the human domination of the physical environment. The "image of God" signifies human uniqueness and distinctiveness, which for Soloveitchik (as for the rabbis) is to be found in self-reflective consciousness. Unlike other animals, the human is spoken to by God and that speech informs him "of his biological propensities and tendencies." The Bible names that self-awareness as the "image of God" (*tzelem*) and it signifies "man's awareness of himself as a biological being and the state of being informed of his natural drives."[78] With the rise of self-awareness begins the ethical consciousness, which renders the human more than a biological reality ("man-nature" in Soloveitchik's terminology). So long as the human was merely part of the natural order, the human experienced God "as revealed in the cosmic processes, in the regularity, unity, and continuity of the same,"[79] which is expressed by the divine name Elohim. With the rise of ethical consciousness, the human experienced God as a personality receiving the commandment not to eat from the tree of knowledge of good and evil (Gen. 2:16–17). With the first commandment, the first ethical norm is established, launching the human experience of "his selfhood, his personalistic existence."[80] As a person ("man-personality" in Soloveitchik's terminology), the human is aware that "his incongruity with the animal was pronounced by the cognitive operation."[81] The shift in awareness marks the existential change from solitude (*badad*) to loneliness (*levad*) to which God responds with the creation of the woman as a counterpoint (*ezer ke-negdo*) to the man. For Soloveitchik, the woman does not denote the material aspect of humanity (as Maimonides understood it) but a partner for dialogical relationship, for an I-Thou fellowship. The ethical is not a lower epistemic state but a higher phase in human development toward freedom.

Complete freedom, according to Soloveitchik, can be found only in the life of halakhah, which for Soloveitchik, as Aviezer Ravitzky explains, is not just "a group of norms guiding our activities but also is a conceptual theoretical structure, addressed toward our cognition."[82] Like the mathematical sciences, halakhah is conceptual, creative, and abstract, but the study of halakhah is superior to the natural sciences, because halakhah is not derived from any concrete, corporeal reality. As a revelation of divine reality, it pertains to the infinite divine intellect understanding of which through the process of halakhic reasoning is the source of endless creativity. It is in the process of legal interpretation that the human becomes "a partner with the Almighty in the act of creation."[83] By virtue of the halakhic life, the life lived through endless choices and creative innovations of legal rulings, the religious personality perpetually creates itself.

Soloveitchik did not develop a full-fledged philosophy of biology or philosophy of science, but his numerous sermons and lectures provide instructive insights about the human condition in regard to science and technology. A thoroughly modern person, Soloveitchik had deep respect for contemporary science and technology and affirmed them as sources of human dignity and human creativity. He believed human scientific endeavors are divinely sanctioned, teaching, "the Creator, as it were, impaired reality in order that mortal man could repair its flaws and perfect it."[84] Indeed, humans are called to be as creative and as majestic as God, and be involved in the cosmos through science and technology and for this reason, Soloveitchik asserts, "Judaism has displayed so much sympathy for scientific medicine and commanded the sick person to seek medical help."[85] But human control and improvement of the physical environment does not exhaust the meaning of being human. To be a partner in God's creation, to truly imitate God, involves not only curing diseases and healing the sick, but also self-creation through interpersonal relationship and the ability to acknowledge defeat and repent one's wrong doing. It is in the process of repentance (*teshuvah*) and the seeking of atonement (*kapparah*) that humans find true creativity, discover purpose and significance, and overcome their sense of loneliness.[86] The true human self-creation is to be found not in enhancement of human biological traits through genetic engineering, but the internal, existential process in which "man ceases to be a mere species man, and becomes a man of God"; in the relationship with God, which is governed by law, the human finds complete freedom.

Leon Kass's cautionary stance toward biotechnology and especially his call for humility can find an interesting ally in the teachings of Soloveitchik. Kass's inspiration, of course, was not the Lithuanian rabbi but the German-Jewish philosopher, Hans Jonas, although Kass's views are not identical with the views of Jonas.[87] Interestingly, Jonas too found the kabbalistic doctrine of divine self-withdrawal (*tzimtzum*) an instructive symbol for his own philosophy of nature and post-Holocaust theology. But if Soloveitchik engaged kabbalah as an integral part of the Jewish normative tradition, Jonas approached kabbalah as an academic scholar who did not observe traditional Judaism. Jonas's philosophy of nature is the most systematic Jewish attempt to address the horrors of the twentieth century brought about by the nihilistic tendencies of modern science.

Hans Jonas: The "Idea of Man" and the Imperative of Responsibility

More than any other Jewish philosopher in the twentieth century, Hans Jonas understood and philosophically addressed the profound impact of modern science and technology. Jonas was a product of a highly assimilated Jewish home in post-Emancipation Germany, even though he had a strong ethnic Jewish identity, forged by secular Zionism.[88] Similarly to Philo in antiquity and Maimonides in the Middle Ages, Jonas immersed himself in contemporary philosophy, engaging it from a Jewish perspective, creatively and critically. Like Soloveitchik, Jonas was trained in German universities, studying with luminaries such as Edmund Husserl (1859–1938), Martin Heidegger (1889–1976), and the Protestant theologian Rudolf Bultmann (1884–1976). Contrary to other German Jews, Jonas did not believe that a cultural synthesis between Germanism (*Deutchtum*) and Judaism, as advocated by Hermann Cohen, was possible and desirable.[89] His groundbreaking dissertation on Gnosticism, written at the University of Marburg in the early 1930s, critically engaged his teacher Heidegger. Even though Jonas employed Heidegger's philosophy throughout his career, he did so while articulating a distinctive Jewish response to Heidegger's nihilism.[90]

Jonas's critical stance toward science and technology is inseparable from his biography. He began his philosophical studies in 1921 as a student in a Jewish institution of higher learning, the Berlin University for the Science of Judaism (*Hochchule für Wissenschaft des Judentums*). In 1928 he studied philosophy at Marburg University along with other Jewish intellectuals who came to study with Heidegger, among them Karl Löwith, Hannah Arendt, and Emmanuel Levinas. With the rise of the Nazis to power, with whom Heidegger fully cooperated as the Rector of Freiburg University, Jonas was forced to flee Germany, first to England in 1933 and two years later to the British Protectorate in Palestine. In 1939, Jonas enlisted in the Jewish Brigade of the British Army and participated in the Italian campaign of 1943. He remained in the British Army until 1945. Soon after his return to Palestine, he once again volunteered for military service and fought in the 1948–1949 War of Independence of the nascent state of Israel. Jonas's combat experience changed his outlook on life and his academic interests. From the history of religion, Jonas turned to the philosophy of nature, extending phenomenology and existentialism to include all forms of life.[91] Unable to secure an academic position in the state of Israel, Jonas migrated once more, first to Canada in 1949 and from there to the United States in 1955, where he joined the faculty of the New School for Social Research.

Unlike the other three philosophers discussed in this chapter, Jonas had no official position in the Jewish community and no Jewish constituency. To this day he remains relatively unknown among Jews, overshadowed by more famous Jewish philosophers and theologians such as Franz Rosenzweig (d. 1929), Martin Buber (d. 1963), Abraham Joshua Heschel (d. 1972), Mordecai Kaplan (d. 1992), Joseph B. Soloveitchik, and Emmanuel Levinas (d. 1995), among others.[92] Nonetheless, Jonas should be counted in the canon of modern Jewish philosophy and be recognized for the originality and depth of his Jewish critique of modern philosophy, science, and technology. A few paragraphs cannot do justice to his rich ideas, but they are necessary if we are to understand how Jewish philosophy may contribute to a cautionary stance toward the new genetics and biotechnology.

To begin, Jonas was keenly aware of the difference between the premodern and modern human condition, between the premodern and modern outlooks.[93] In the premodern world, as Philo and Maimonides illustrate, the human condition was believed to be inherent in the very nature of things; the human good was determinable on that basis, and human action and responsibility were narrowly circumscribed.[94] All this changed radically in the modern world as a result of modern technology, which has amply demonstrated that human action can radically alter nature, including human nature. While nature has never been immune to human intrusion, modern biotechnology has made it possible to expand the human life span, control human behavior, and manipulate the genome, all of which raise questions about what is normatively human while putting human nature itself at stake.[95]

What is unique about modern technology is that it sets in motion a causal chain that has profound effect on objects and peoples in very remote places and in future epochs, and that these changes are *irreversible*. If mistakes are made, correcting them is very difficult, and in many cases impossible. Therefore, modern technology, especially biotechnology, changed the moral situation and undermined the entire framework of human action. Modern technology is inherently utopian and progress is an inherent drive of technology. This utopian vision is behind the promises of technology to cure the "mistakes" of nature, or overcome its shortcoming. As a result of the new advances in molecular biology human beings have become the very object of technology. As Jonas put it, "*Homo faber* [man the maker], is now turning on himself and gets ready to make over the maker of all the rest."[96] According to Jonas, the modern dynamics of progress and the material emphasis on *making* and *fabricating* culminates in the remaking of the maker that characterizes modern technology. Modern technology, including biomedicine, keeps moving forward incessantly, threatening the very source of the technological project, the human. Since we now have the power to alter or destroy humanity itself, Jonas argues that the existence of humanity as created by God has become a value. Our genetic engineering means that we are now able to create other humans, not in the image of God, but in our own image. This is the hubris against which Jonas spoke up with prophetic passion.

Jonas's second and related insight is that technology is more than a mere instrument for human purposes. Modern technology forces human beings into a dialectical situation: our attempt to bring the external world under our power ends with the power of technology to destroy or radically refashion the very subject whose power it is. In the premodern world technology was not able to alter nature and remained ethically neutral because it did not have the power to fundamentally alter or destroy nature; *techné* was not "the road to mankind's chosen goal." Instead, as we have seen in the case of Maimonides, the goal of human life was contemplation (*theoria*), which was far from being ethically neutral; rather its object was the attainment of the highest good. To act against the good was therefore to act in contradiction to knowledge. In the modern world the contemplation of things is not considered the goal of humanity; the chosen goal is having *power over things*. It means "elevating *homo faber* to the essential aspect of man. At its most extravagant, it means elevating power to the position of his dominant and indeterminable goal."[97] Thus modern technology lost its ethical neutrality and is no longer innocent. Even the pursuit of knowledge has changed because it too has been reduced to technology; knowledge is no longer about the good. Ethically neutral, knowledge is now used for good or evil, and experts, namely the scientists, do not have the capacity to determine the proper and improper uses of their knowledge. The only way to address this conundrum is to change the way we understand the relationship between humanity and the natural world. Jonas offers a new philosophy of nature that recognizes the intrinsic moral value of the natural world and our responsibility toward the natural world. Accordingly, he argues, the natural world is not just inert material stuff that we are free to do with what we please, rather, it has an inherent moral significance. Applying Kantian principles and distinctions to nature, Jonas insists that nature should be treated as an end and not merely as a means.

The third insight follows from the first two. Jonas was the first Jewish thinker to take note of the ecological crisis due to developments in modern technology and to think philosophically about genetic engineering. He understood the capacity to refashion human life through genetic, biochemical, and neurological interventions and was the first to caution humanity to exercise a "heuristics of fear" vis-à-vis technology.[98] Jonas saw himself as a champion of organic life, precisely because he personally experienced the vast devastation of life. For Jonas, organic life is itself "an ontological revolution in the history of matter," a radical change in matter's mode of being. He gave a phenomenological account of organic life, including humanly organic life. Beginning with the most basic phenomenon of metabolism, Jonas interprets all organic life and individual organisms in nondualistic terms. His concern was to overcome the radical split between nature and ethics, between "is" and "ought," characteristic of modern science and philosophy. This will be achieved, Jonas thought, by endowing all forms of life with intrinsic moral worth and by insisting that life and the material world, necessary for its being, command ultimate respect, allegiance, and finally moral commitment.

At the center of Jonas's philosophy of life is the notion of responsibility, a very Jewish notion indeed, which he understood to mean "responsibility for something" as well as "responsibility to something—to an ultimate authority to which an accounting must be given."[99] He distinguished between "formal responsibility," which is neutral, and "substantive responsibility," which is "responsibility for the caring or preservation of some object." Jonas did not use religious arguments in the service of public ethical debates; he only set out to articulate the wisdom that can be gleaned from the Jewish stance on the relation between power and responsibility. Focusing on the Bible rather than on rabbinic sources, Jonas insisted that the Torah teaches not only respect of nature but also respect for our human nature. Creation in the "image of God" means the ability to distinguish between good and evil and our responsibility for promoting the good, symbolized by the commandments. Jonas is thus in agreement with Philo and Maimonides (whom he cites) that we should seek the improvement of character through education, although he comes to this conclusion from a different point of departure. For Jonas the focus on cultivation of character is rooted in the respect for the mystery of human freedom, and it is this freedom that we must protect by preferring persuasion under conditions of freedom to psychological manipulation in the hands of behavioral engineers. According to Jonas, then, medicine should not be transformed into the effort to eliminate imperfection or to prolong life at all cost.

Jonas's ethics of responsibility was future directed, encompassing responsibility toward future generations to ensure the future of humanity. Such an ethics develops a conception of humanity not as it is, but as it is yet to be realized, very much as Philo and Maimonides did before him. For Jonas the very existence of humanity is an objective good that imposes an obligation on the human will, which through technology has power over this objective good. Jonas asserts: "There is an unconditional duty for mankind to exist."[100] The source of the duty is an "ought" that stands above both ourselves and future human beings. It is "their duty over which we have to watch, namely, their duty to be truly human; thus over their capacity for this duty . . . which we could possibly rob them of with the alchemy of our 'utopian' technology."[101] Jonas summarizing the point says, "No condition of future descendents of humankind should be permitted to arise which contradicts the reason why the existence of mankind is mandatory at all."[102] He further explains, "With this imperative, we are, strictly speaking, not responsible to the future human individuals but to the idea of Man. . . . It is this ontological imperative, emanating from the idea of Man, which stands behind the prohibition of a gamble with mankind. Only the *idea of Man*, by telling us *why* there should be men, tells us also *how* they should be."[103] The very idea of humanity makes the existence of humanity an imperative; *humanity ought to be*. It will thereby set a limit to those technologies that in their quest for control over necessity and improvement of human nature threaten the existence of humanity so conceived.

It is easy to see that what Jonas calls the "Idea of Man," or the "idea of humanity" is what premodern Jewish philosophers defined as "the image of God." This similarity is not grounded in speculation, since Jonas reflected specifically on the biblical narratives of creation. Jonas imagines that God pronounces creation to be good only with the long-awaited but accidental emergence of life: of creatures who value their own existence against the threat of death. Prior to the advent of knowledge, God's cause cannot go wrong because life retains its innocence. Eventually with the evolution of humanity, life arrives at the highest intensification of its own value. For our capacities for knowledge and freedom represent "transcendence awakened to itself." The price is that with knowledge and freedom come the power to will and do evil, and this power becomes absolute in a technological age with our ability to destroy ourselves. Moral responsibility is the mark of our being made "for"

God's image, though not "in" it.[104] Among earthly creatures only we can acknowledge the transcendent importance of our deeds: that we are "mortal trustees of an immortal cause."

To God's self-limitation we owe the existence of things, Jonas teaches in agreement with Lurianic kabbalah, for God's withdrawal gives us human beings room to help Him by bearing responsibility for our own vulnerable affairs.[105] The immortality of deeds is for God's sake not for human sake. One must care about the divine in its own right in order for it to bear upon how we spend our time. But one has reason to care about God's destiny if one experiences God's goodness reflected in the immanent goodness of creation itself. Thus Jonas answers the question "why ought we to worry about the distant future of humankind and the planet" only on the basis of comprehensive ontology without recourse to theology. He maintains that even if there is no Creator, and instead the world has always been or came into being mindlessly, and even if there is no God to ensure the immortality of our deeds, Being remains a good in itself, in virtue of the presence of purposiveness within it. In departure from the premodern belief in the afterlife, Jonas did not believe in immortality of souls. He admits that the premodern view conflicts with modern science, but he affirms human mortality and the immanence of nature in accord with the modern scientific temper while showing its limitations.

In contrast to the mechanistic worldview that characterizes the modern project, Jonas shows the purposiveness of nature and in this regard he is very much in accord with premodern thinkers, especially the Aristotelian tradition of which Maimonides was the most notable Jewish example. Jonas goes beyond the Aristotelian teleological framework when he provides a phenomenology of life that is rooted in the lowest functions of the living organism—metabolism.[106] Metabolism shows that organisms are not machines; in metabolism the organism both is and is not identical to its matter; it consists of matter but it is not identified with matter, because then it would be dead. The identity of the organism consists in performance or self-integration, which implies purposiveness. Nature then has purposes and therefore has value.

How do Jonas's philosophical insights shape attitudes toward biotechnology? In terms of life expansion, Jonas reminded us that mortality is not just a curse or a burden, it is also a blessing.[107] It is a burden insofar as we are organic beings, we must wrest our being from the continuous threat of nonbeing. But it is a blessing insofar as our wresting is the very condition for any affirmation of being at all, so that "mortality is the narrow gate through which alone value—the addressee of a yes—could enter the otherwise indifferent universe."[108] For Jonas, the effort to forestall death or even mortality itself is a fundamental denial of what makes us human. The process of life requires mortality as the counterpart of the natality that alone can supply the novelty and creativity that enrich human life and express freedom. Freedom is imperiled when it ignores necessity. In terms of genetic engineering Jonas considered many ends of genetic engineering to be frivolous. Genetic enhancement for the sake of improving one's look or one's chances of social success falls into that category. As for germline intervention, here he appeals exclusively to consequences: the irreversibility of germline interventions, the range of their effects, the impossibility of drawing a line in practice between therapy and enhancement of traits or prohibiting the outright invention of new human forms that isolate the ontological states of human nature. In terms of human improvement, or eugenics, Jonas distinguished between *negative eugenics* (namely, developing diagnostic tools to identify genetic diseases and then manipulating the genetic code to eliminate bad genes) and *positive eugenics* (namely, manipulating genes so as to enhance human performance). In regard to both programs he reminds us that an ambitious eugenics violates the normative status of nature, but also that we do not have the criteria or standards to determine what is normal and what is pathogenic. As for elimination of "bad genes" from the population, any effort to eliminate undesirable genes from the gene pool altogether threatens the biological necessity of a varied gene pool and encounters our ignorance about the role apparently useless genes may play in human adaptability. Jonas argues against positive eugenics on the same ground. The lack of criteria and standards for intervention means that positive eugenics that aims at a qualitative improvement over nature cannot claim the sanction of nature.

Conclusion

Judaism does not speak in one voice on any issue, including the creation of man in the image of God and the application of this theological construct to contemporary biotechnology. However, the reflections of Philo, Maimonides, Soloveitchik, and Jonas offer us ways to think philosophically about the new genetics and its corollary biotechnology and not just halakhically or medically. Contrary to the enthusiastic endorsement of the new genetics and biotechnology, the Jewish philosophical tradition compels us to ask: What is the vision of the human presupposed by biotechnology? Does this vision fit with our Jewish moral sensibilities? What is it that we seek to improve and what counts as human improvement? Whose values will determine that a given technology is good? Who will possess the power to make these decisions? How will these decisions cohere with our commitment to democracy and the value of human dignity and nobility? While the four Jewish philosophers represent different historical epochs and philosophical schools, they also share important beliefs about the world and the place of human beings in it. The world is created by God and not by human beings; the created world is purposeful and its created structure contains ethical norms; humans who are endowed with rationality have the moral responsibility to protect the world created by God; *they were not given a license to re-create the world in their own image.* In their own distinctive way, Philo, Maimonides, Soloveitchik, and Jonas compel us not to forget who we are: created beings who have much in common with all other forms of life, but who are also unique and different from other living creatures. As humans, we are neither merely animals nor are we machines, with replaceable internal parts. The symbol of the divine "image" expresses our uniqueness, indicating that to reduce humans either to mere forms of life or to machines is morally, religiously, and intellectually unacceptable.[109] We will continue to explore and break new grounds in genetic technologies, but I for one believe that each phase of development should be subject to intense scrutiny in light of the impact of the technology on all aspects of life. Religious and secular Jews who have enthusiastically embraced the new genetics and its corollary biotechnology in order to solve the problem of physical, Jewish survival, should be aware of the cautionary stance articulated by Jewish philosophers. As serious as the current demographic crisis us, it is doubtful that the solution should be framed in strict somatic or material terms.

Humanity today is on the verge of a new phase in its development that some have called "posthumanism" or "transhumanism."[110] Transhumanists deny the uniqueness of human beings, considering it not only as "speciesism" but as "human racism"; they eliminate boundaries between humans and machines and improperly give the status of "person" to animals, cyborgs, and intelligent machines. While the philosophical merits of transhumanism are yet to be evaluated in depth, the transhuman future as envisioned by its proponents should give us all cause for serious concern. The fact that critics of transhumanism, such as, Leon Kass, Francis Fukuyama, William Kristol, Jean Bethke Elshtain, George Annas, and Bill McKibben come from diverse and even conflicting religious and political perspectives suggests that the critics, even better than the proponents, correctly understand what is at stake in the new advances of science and technology—the very existence of humanity now hangs in the balance and we must proceed, therefore, with great caution.

Jews have reflected on these challenges within the matrix of the halakhic process and have been generally supportive of biotechnology, dismissing critics such as Kass with ad hominem arguments. Yet, as this essay attempts to show, the critique of the new genetics is neither silly nor shallow; rather, it speaks about human dignity from within the ancient tradition of ethical virtue to which Aristotle and his Jewish followers, especially Philo and Maimonides, gave the most succinct expression. Ethical virtue is very much part of Judaism and preserving it for future generations is necessary for Jewish spiritual survival. Yes, we humans are embodied beings but we are not machines whose declining parts can be replaced indefinitely; we are more than the sums of our organic parts and that "more" is referred to by the Jewish tradition as the "image of God." The Jewish philosophical tradition reminds us that Jewish existence is not merely a matter of birth and shared genetics; it involves no less a commitment to a Torah-centered life, however one interprets the meaning of "Torah." The human body is indeed the foundation of continued Jewish existence, but the perpetuation of the body alone is not the *telos* of Jewish existence. Jewishness encompasses no less a nonbiological dimension shaped by commitment to Torah, however defined and interpreted. Following Philo, Maimonides, Soloveitchik, and Jonas contemporary Jews could and should engage the Jewish tradition philosophically and wrestle with the challenges posed by the new genetics and its accompanying biotechnology.

Notes

1. For an excellent overview of the relevant issues posed by the new genetics and its corollary biotechnology consult Audrey R. Chapman, *Unprecedented Choices: Religious Ethics at the Frontiers of Genetic Science* (Minneapolis: Augsburg Press, 1989) and the extensive bibliography cited there.

2. Leon R. Kass, *Life, Liberty and the Defense of Dignity: The Challenges for Bioethics* (San Francisco: Encounter Books, 2003), 10.

3. Leon R. Kass, ed., *Human Cloning and Human Dignity: The Report of the President's Council on Bioethics* (New York: Public Affairs, 2002).

4. See Leon R. Kass, *The Beginning of Wisdom: Reading Genesis* (New York: Free Press, 2003). In *Life, Liberty and the Defense of Dignity*, Kass's Jewish identity is stated most explicitly in the last chapter, "*L'Chaim* and Its Limits: Why Not Immortality?" 257–274.

5. Kass, *Life, Liberty and the Defense of Dignity*, 258.

6. Some of the most cited works by believing Christians include: Robert Song, *Human Genetics: Fabricating the Future* (Cleveland, Ohio: The Pilgrim Press, 2002); Ron Cole-Turner; Maureen Junker-Kenny and Lisa Sowle Cahill, eds., *The Ethics of Genetics Engineering* (London: SCM Press; and Maryknoll, N.Y.: Orbis, 1996); Ted Peters, *Playing God* (London: Routledge, 1997); Celia Dean-Drummond, *Biology and Theology Today: Exploring the Boundaries* (London: SCM Press, 2001); and Dean-Drummond, *Creation Through Wisdom: Theology and the New Biology* (London: T and T Clark, 2000).

To date the most comprehensive survey of Jewish attitudes toward the new genetics is Miryam Z. Wahrman, *Brave New Judaism: When Science and Scripture Collide* (Lebanon, N.H.: University Press of New England for Brandeis University Press, 2002).

7. Rabbi J. David Bleich, "Cloning: Homologous Reproduction and Jewish Law," *Tradition* 32 (1998): 47–86. Rabbi Bleich tends to rule more stringently than other Orthodox jurists. For his position on a host of related legal issues consult his *Bioethical Dilemmas* (Hoboken, N.J.: Ktav Publishing House, 1998); *Judaism and Healing: Halakhic Perspectives* (Hoboken, N.J.: Ktav Publishing House, 1981).

8. Fred Rosner, "Recombinant DNA, Cloning and Genetic Engineering in Judaism," *New York State Journal of Medicine* 79 (1979), 1442; cited in Wahrman, *Brave New Judaism*, 71. Rosner's position on a variety of medical issues are to be found in his *Modern Medicine and Jewish Ethics* (Hoboken, N.J.: Ktav Publishing House; New York: Yeshivah University Press, 1986).

9. Babylonian Talmud, Niddah 31a; Kiddushin 30b, Shabbat 10a.

10. Azriel Rosenfeld, "Judaism and Gene Design," *Tradition* 13 (1972): 71–80.

11. Whether the Talmud should be understood to mean that the human is a cocreator is open for discussion. The notion of the human as divine cocreator was articulated most fully in Christian theology by Philip Hefner, *The Human Factor: Evolution, Culture and Religion* (Minneapolis: Fortress Press, 1993).

12. Elliot N. Dorff, *Matters of Life and Death: A Jewish Approach to Modern Medical Ethics* (Philadelphia: Jewish Publication Society, 2003), 157. The reference to human partnership with God as a justification for a pro-biotechnology stance is developed in Byron Sherwin, *In Partnership with God* (Syracuse, N.Y.: Syracuse University Press, 1990). Later in the chapter I discuss how Rabbi Joseph B. Soloveitchik understood human partnership with God, which leads to a much more cautious attitude toward biotechnology.

13. Dorff, *Matters of Life and Death*, 157. Emphasis added.

14. Ibid., 321.

15. Ibid., 322; cf., Dorff, "Human Cloning: A Jewish Perspective," Presentation to the National Bioethics Advisory Commission, March 14, 1997, 6. The same position is also endorsed by Rabbi Byron Sherwin, "Religious Perspectives on Cloning: A Jewish View," Presentation at the U.S. Capitol, June 24, 1997.

16. Barbara Prainsack and Ofer Firestine, "'Science for Survival': Biotechnology Regulation in Israel," *Science and Public Policy* (February 2006): 34.

17. The primacy of science in the advancement of national goals was one of the marks of secular Zionism that brought about the creation of the state of Israel. For a fuller analysis of the use of science in the Zionist movement, see Derek J. Penslar, *Zionism and Technocracy: The Engineering of Zionist Settlement in Palestine, 1870–1918* (Bloomington and Indianapolis: Indiana University Press, 1991).

18. Cited in Wahrman, *Brave New Judaism*, 68.

19. See Richard M. Goodman, *Genetic Disorders among the Jewish People* (Baltimore: Johns Hopkins University Press, 1979); Richard M. Goodman and Arno G. Motulsky, *Genetic Diseases among Ashkenazi Jews* (New York: Raven Press, 1979); Batsheva Bonné-Tamir and Avinoam Adam, eds., *Genetic Diversity among the Jews* (New York: Oxford University Press, 1992).

20. On recent trends of Jewish demography consult Egon Mayer, Barry A. Kosmin, and Ariela Keysar, *American Jewish Identity Survey 2001: AJIS Report: An Exploration in the Demography and Outlook of a People* (New York: Center for Cultural Judaism, 2003).

21. According to Sofer's analysis, within the pre-1967 borders the development is similarly pessimistic for the Jewish majority: the current population of more than 5

million Jews and 1.2 million Arabs will change to a ratio of 6.6 million Jews to 2.1 million Arabs (Muslim, Christian, and Druze) in Israel proper.

22. See Susan Martha Kahn, "Rabbis and Reproduction: The Uses of New Reproductive Technologies among Ultraorthodox Jews in Israel," in *Infertility around the Globe: New Thinking on Childlessness, Gender and Reproductive Technologies*, ed. Marcia C. Inhorn and Frank Van Balen (Berkeley: University of California Press, 2002), 283–297; Susan Martha Kahn, *Reproducing Jews: A Cultural Account of Assisted Conception in Israel* (Durham, N.C., and London: Duke University Press, 2000); Jacqueline Portuguese, *Fertility Policy in Israel: The Politics of Religion, Gender and Nation* (Westport, Conn.: Praeger, 1998).

23. Susan Martha Kahn, "Are Genes Jewish? Conceptual Ambiguities in the New Genetic Age," David W. Belin Lecture in American Jewish Affairs (Ann Arbor: Jean and Samuel Frankel Center for Judaic Studies: University of Michigan, 2005), 10. For an elaborate discussion of these issues see Susan Martha Kahn, *Reproducing Jews.*

24. Susan Martha Kahn, "Are Genes Jewish?" 10.

25. The most sustained argument for this position is offered by Michael Wyschogrod, *The Body of Faith: God and the People of Israel* (Northvale, N.J.: Jason Aronson, 1996, and *Abraham's Promise: Judaism and Jewish-Christian Relations*, ed. R. Kendall Soulen (Grand Rapids, Mich.: W. B. Eerdmans / London: SCM Press, 2004).

26. For an overview of the history of interpretation see Alexander Altmann, "Homo Imago Dei in Jewish and Christian Theology," *Journal of Religion* 48 (1968): 235–259.

27. See Harry A. Wolfson, *Philo: Foundation of Religious Philosophy in Judaism, Christianity and Islam* (Cambridge: Harvard University Press, 1947).

28. David T. Runia, *Philo in Early Christian Literature: A Survey on Philo in the Christian Church* (Assen: Van Gorcum; Minneapolis: Fortress Press, 1993).

29. See Byron L. Sherwin, *In Partnership with God: Contemporary Jewish Law and Ethics* (Syracuse: Syracuse University Press, 1990); *Jewish Ethics for the Twenty-First Century: Living in the Image of God* (Syracuse: Syracuse University Press, 2000).

30. For a good overview of Philo's life, career, and philosophy consult Ronald Williamson, *Jews in the Hellenistic World: Philo* (Cambridge: Cambridge University Press, 1989).

31. The cultural and religious context of Philo's doctrine is discussed at length in Hindi Najman, "The Symbolic Significance of Writing in Ancient Judaism," in *The Idea of Biblical Interpretation: Essays in Honor of James L. Kugel*, ed. Hindi Najman and Judith H. Newman (Leiden, Boston: Brill, 2004), 139–173; Hindi Najman, *Seconding Sinai: The Development of Mosaic Discourses in Second Temple Judaism* (Leiden, Boston: Brill, 2005), 70–107.

32. For an exposition of Philo's view of the ultimate end of human life consult Hava Tirosh-Samuelson, *Happiness in Premodern Judaism: Virtue, Knowledge and Well-Being* (Cincinnati: Hebrew Union College Press, 2003), 81–100 and the sources cited there.

33. Philo composed a commentary on the *Timaeus*. For a critical edition and analysis of this text see David T. Runia, *Philo of Alexandria and the Timaeus of Plato* (Leiden: E.J. Brill, 1986).

34. Philo, *De Opificio* 4, 16

35. Logos is an *eikon* of God that is be found in the rational order visible in the universe. That order is accessible to human beings because they are created in the "image of God," namely with a reasoning power, the soul.

36. Philo, *De Opificio*, 135.

37. Ibid., 146.

38. It is most likely that Philo's association between the "seal" and the "coin" reflects the cult of the Roman emperors in late antiquity. See Yair Lorberbaum, *Image of God: Halakha and Aggadah* (in Hebrew) (Tel Aviv: Schocken, 2004), 307–308.

39. An example of this view is Alon Goshen-Gottstein, "The Body as Image of God in Rabbinic Literature," *Harvard Theological Review* 87:2 (1994): 171–195. He calls for "the liberation of rabbinic theology from the reign of medieval theology" (that is, from the reading proposed by Maimonides and Philo before him). According to Goshen-Gottstein the rabbis developed two interpretations of the construct "image of God": according to one interpretation, the image refers to a "luminous entity" whose brightness was diminished when the first human sinned so that afterward the rabbis did not consider the human as created in the divine image. According to the second interpretation, the "image" referred to bodily shape and it is continuous and unchanging since the creation of the first human. For Goshen-Gottstein, the first reading relates to the legal rulings of the rabbis, whereas the second interpretation has only didactic usage. Even if Goshen-Gottstein's reading of rabbinic sources is correct, it does not invalidate the merit of Philo's and Maimonides' views that I discuss in this chapter. Lorberbaum's study also discusses the rabbinic traditions that understand the "image" in an embodied fashion but his conclusions are very different from Goshen-Gottstein because the "image" in rabbinic thought consists of more than the somatic body to include "personality, consciousness, and sensations, in short, all the mental components." See Lorberbaum, *Image of God*, 333.

40. Babylonian Talmud, Berachot 10a: "There is none holy as the Lord, for there is none besides You, 'neither is there any rock like our God' [I Samuel 2:2]. What means 'neither is there any rock [*tzur*] like our God'? There is no artist [*tzayyar*] like our God."

41. For this insight I am indebted to John F. Kavanaugh, S.J., *Who Counts as Persons?: Human Identity and the Ethics of Killing* (Washington, D.C.: Georgetown University Press, 2001).

42. The essay by Alon Goshen-Gottstein cited earlier is an example of such a trend.

43. On Philo's notion of human stewardship consult Katell Berthdelot, "Philo and Kindness toward Animals (*De Virtutibus* 125–147)," *The Studia Philonica Annual* 14 (2002): 48–65.

44. The best exposition of Maimonides' life, works, and literary sources is offered by Herbert A. Davidson, *Moses Maimonides: The Man and His Works* (New York: Oxford University Press, 2005).

45. A good exposition of Maimonides' nonembodied understanding of God is offered in Kenneth Seeskin, *Searching for a Distant God: The Legacy of Maimonides* (New York: Oxford University Press, 2000).

46. *Shiur Qomah* literally means "The Stature of the [Divine] Body" and the term denotes a literary corpus that consisted of speculations about the measurements of God's body expressed in astronomical units. Whether these speculations emerged originally within rabbinic circles or in circles antagonistic to the rabbis is a hotly debated issue in current scholarship. Sometime between the eighth and ninth centuries, these speculations were absorbed into the rabbinic tradition and became the foundation of further mystical and theosophic speculations among the medieval kabbalists. For an overview of this tradition see Joseph Dan, "The Religious Experience of the Merkavah," in *Jewish Spirituality from the Bible through the Middle Ages*, ed. Arthur Green (New York: Cross Road, 1986), 287–307.

47. For a good exposition of Maimonides' interpretation of the first human sin consult Lawrence Berman, "On the Fall of Man," *AJS Review* 5 (1980): 1–15.

48. See Hava Tirosh-Samuelson, "Gender and the Pursuit of Happiness in Maimonides' Philosophy," in *The House of Jacob—Images and Paradigms: Gender and Family in Medieval Jewish Society*, ed. Elisheva Baumgarten, Roni Weinstein, and Amnon Raz-Karkotzkin (Jerusalem: Bialik Institute, forthcoming); Susan Shapiro, "A Matter of Discipline: Reading for Gender in Jewish Philosophy," in *Judaism since Gender*, ed. Miriam Peskowitz and Laura Levitt (New York and London: Routledge, 1997), 158–173.

49. Babylonian Talmud Shabbat 133b: "This is my God and I will glorify Him; Abba Saul said: 'Be like (*domeh*) Him, as He is merciful so shall you be merciful . . .'"

50. For further exposition of this point consult Hava Tirosh-Samuelson, *Happiness in Premodern Judaism*, 223–231.

51. Ibid., 206–214.

52. One notable exception to this generalization is Walter S. Würzburger's *Ethics of Responsibility: Approaches to Covenantal Ethics* (Philadelphia: Jewish Publication Society, 1994) that argues that Jewish law includes an understanding of ethical virtue derived from the commandment of *imitation Dei* (Deut. 29:9). For a review of the book consult David Shatz, "Beyond Obedience: Walter Würzburger's *Ethics of Responsibility*," *Tradition* 30:2 (1996): 74–95. For elaboration of Würzburger's approach see Yitzchak Blau, "The Implications of Jewish Virtue Ethics," *The Torah U-Madda Journal* 9 (2000): 19–41.

53. See Daniel H. Frank, "Humility as a Virtue: A Maimonidean Critique of Aristotle's Ethics," in *Moses Maimonides and His Time*, ed. Eric Ormsby (Washington, D.C.: The Catholic University of American Press, 1989), 88–99.

54. The Maimonidean Controversy engulfed world Jewry throughout the thirteenth century and continued to erupt periodically, for example, in the sixteenth cen-

tury. The controversies about Maimonides encompassed social, political, cultural, and personal issues and they ultimately pertained to the meaning of being human and the nature of human happiness. See Hava Tirosh-Samuelson, *Happiness in Premodern Judaism*, 246–260, 563–564.

55. See Hava Tirosh-Samuelson, "Philosophy and Kabbalah, 1200–1600," *Cambridge Companion of Medieval Jewish Philosophy*, ed. Daniel H. Frank (Cambridge: Cambridge University Press, 2000), 218–257.

56. See Gershom Scholem, "*Tselem*: The Concept of the Astral Body," in his *The Mystical Shape of the Godhead: Basic Concepts in the Kabbalah* (New York: Schocken Books, 1991), 251–273.

57. A useful overview of kabbalistic anthropology is offered in Moshe Hallamish, *An Introduction to the Kabbalah*, trans. Ruth Bar-Ilan and Ora Wiskind-Elper (Albany: State University of New York Press, 1999), 121–166.

58. For a full exposition of this notion consult Elliot Wolfson, "Mirror of Nature Reflected in the Symbolism of Medieval Kabalah," in *Judaism and Ecology: Created World and Revealed Word*, ed. Hava Tirosh-Samuelson (Cambridge: Harvard University Press, 2002), 305–332; Hava Tirosh-Samuelson, "The Textualization of Jewish Mysticism," in *Judaism and Ecology*, 389–404.

59. On the golem tradition see Gershom Scholem, *On Kabbalah and Its Symbolism* (New York: Schocken, 1965), 158–204; Moshe Idel, *Golem: Jewish Magical and Mystical Traditions on the Artificial Anthropoid* (Albany: State University of New York Press, 1990); Byron Sherwin, *The Golem Legend: Origins and Implications* (Lanham, Md.: University Press of America, 1985). Sherwin finds in the golem tradition the inspiration for his pro-biotechnology stance. Byron L. Sherwin, *Golems among Us: How a Jewish Legend Can Help Us Navigate the Biotech Century* (Chicago: Ivan R. Dee, 2004). The question whether a clone should be considered a golem is considered by Orthodox thinkers such as Rabbi J. David Bleich, who highlights the difference between the two. See Wahrman, *Brave New Judaism*, 73–74.

60. On court of Rudolph II in Prague and the contact between Jewish and non-Jewish scholars consult Noah Efron, "Irenism and Natural Philosophy in Rudolfine Prague: The Case of David Gans," *Science in Context* 10: 4 (1997): 627–649.

61. For an excellent short exposition of German Orthodoxy and its affinity with Kantianism see David Ellenson, "German Jewish Orthodoxy: Tradition in the Context of Culture," in *The Uses of Tradition, Jewish Continuity in the Modern Era*, ed. Jack Wertheimer (New York and Jerusalem: The Jewish Theological Seminary, 1992), 5–22.

62. The depth and breadth of Soloveitchik's philosophical training are evident in his *Halakhic Mind* (London: Seth Press, 1986). For an exposition of Soloveitchik's engagement with Natorp, Cohen, and modern science consult "Towards a Genuine Jewish Philosophy: *Halakhic Mind*'s New Philosophy of Religion," in *Exploring the Thought of Rabbi Joseph B. Soloveitchik*, ed. Marc D. Angel (Hoboken, N.J.: Ktav Publishing House, 1997), 179–206.

63. For example, Shubert Spero and Walter S. Würzburger present Soloveitchik as a Maimonidean thinker, whereas Moshe Sokol argues that Soloveitchik did not

intend to accomplish the harmonization of halakhah and philosophy worked out by Maimonides. See Shubert Spero, "Rabbi Joseph Dov Soloveitchik and the Role of the Ethics," *Modern Judaism* 23 (2003): 12–31; Walter S. Würzburger, "The Maimonidean Matrix of Rabbi Soloveitchik's Two-Tiered Ethics," in *Through the Sound of Many Voices*, ed. J. V. Plant (Toronto: Lester and Orpen Dennys, 1982), 172–183; Moshe Sokol, "Ger Ve-Toshav Anokhi: Modernity and Traditionalism in the Life and Thought of Rabbi Joseph B. Soloveitchik, in *Engaging Modernity: Rabbinic Leaders and the Challenge of the Twentieth Century* (Northvale, N.J., and Jerusalem: Jason Aronson, 1997), 149–165. The essays by Würtzburger and Sokol are reprinted in *Exploring the Thought of Rabbi Joseph B. Soloveitchik*," ed. Mark D. Angel (Hoboken, N.J.: Ktav Publishing House, 1997). The best exposition of Soloveitchik's relationship to Maimonides' philosophy is offered by Aviezer Ravtizky, "Rabbi J.B. Soloveitchik on Human Knowledge: Between Maimonidean and New-Kantian Philosophy," *Modern Judaism* 6:2 (1986): 157–188.

64. The point is made by David Hartman, *Love and Terror in the God Encounter: The Theological Legacy of Rabbi Joseph B. Soloveitchik* (Woodstock, Vt.: Jewish Lights, 2001).

65. See Soloveitchik, "Majesty and Humility," *Tradition* 17:2 (1978): 25–37. The essay was delivered originally as a lecture at Rutgers University on April 14, 1973.

66. Both Moshe Sokol and Walter S. Würzburger highlight the personal biography of Soloveitchik whereas David Hartman minimizes it. See Sokol's essay cited earlier and Walter S. Würzburger, "Rav Soloveitchik as a Posek of Postmodern Orthodoxy," in *Engaging Modernity*, 123.

67. Soloveitchik, "Majesty and Humility," 33.

68. For a useful exposition of the doctrine consult Gershom Scholem, *Kabbalah* (Jerusalem: Keter, 1974), 128–140.

69. Soloveitchik, "Majesty and Humility," 36.

70. Structurally speaking Soloveitchik's typology is reminiscent of Augustine's (and perhaps was inspired by it) even though Soloveitchik developed his typology of the Halakhic Man as a critique of and response to the Christian *homo religiosus*.

71. Already in the Middle Ages Jewish philosophers such as Abraham Shalom (d. ca. 1492) noted that the numerical value of the name Elohim is identical with the numerical value of the word "*hateva*" (meaning, "the nature"). Both philosophers and kabbalists understood the name Elohim to refer to the cosmic aspect of God, long before Spinoza articulated his monistic philosophy of *Deus sive natura*.

72. Hartman, *Love and Terror*, 107.

73. Soloveitchik, "Lonely Man of Faith," *Tradition* 7:2 (1965): 56.

74. Hartman, *Love and Terror*, 120.

75. Soloveitchik, *The Emergence of Ethical Man*, ed. Michael S. Berger (Jersey City: Ktav Publishing House, 2005).

76. Ibid., 44.

77. Ibid., 56.

78. Ibid., 76.

79. Ibid., 76.

80. Ibid., 88.

81. Ibid., 91.

82. Aviezer Ravitzky, "Rabbi J. B. Soloveitchik, on Human Knowledge: Between Maimonidean and New-Kantian Philosophy," *Modern Judaism* 6:2 (1986): 160.

83. Soloveitchik, *Halakhic Man*, trans. Lawrence Kaplan (Philadelphia: Jewish Publication Society, 1983), 81.

84. Ibid., 101.

85. Soloveitchik, "Majesty and Humility," 34.

86. For exposition of this consult Pinchas Hacohen Peli, "Repentant Man—A High Level in Rabbi Joseph B. Soloveitchik's Typology of Man," in *Exploring the Thought of Rabbi Joseph B. Soloveitchik*, 229–259.

87. Kass describes his personal association with Jonas in his "Appreciating *The Phenomenon of Life*," *Hastings Center Report* 25:7 (1995): 3–12.

The differences between Jonas and Kass are explained most clearly by Lawrence Vogel, "Natural Law Judaism? The Genesis of Bioethics in Hans Jonas, Leo Strauss, and Leon Kass," *Hastings Center Report* (May–June 2006): 32–44. This essay is reprinted in Hava Tirosh-Samuelson and Christian Wiese, eds., *The Legacy of Hans Jonas: Judaism and the Phenomenon of Life* (Leiden and Boston: Brill Academic Publishers, 2008), 287–314.

88. For a biography of Jonas consult Christian Wiese, *The Life and Thought of Hans Jonas: Jewish Dimensions* (Waltham, Mass.: Brandeis University Press, 2007). A useful overview is also available in Richard Wolin, *Heidegger's Children: Hannah Arendt, Karl Löwith, Hans Jonas, and Herbert Marcuse* (Princeton: Princeton University Press, 2001), 101–133. A good summary of Jonas's philosophy is provided by Lawrence Vogel, "Hans Jonas's Exodus: From German Existentialism to Post-Holocaust Theology," in *Mortality and Morality: A Search for the Good after Auschwitz*, ed. Lawrence Vogel (Evanston, Ill.: Northwestern University Press, 1996), 1–40. New studies on Jonas continue to appear as his significance for the present conundrum is being recognized. For a comprehensive overview consult Tirosh-Samuelson and Wiese, eds., *The Legacy of Hans Jonas* and the comprehensive bibliography (see 412–553).

89. See Steven Schwarzschild, "'Germanism and Judaism': H. Cohen's Normative Paradigm of the German-Jewish Symbiosis," in *Jews and Germans from 1860 to 1933: The Problematic Symbiosis*, ed. David Bronsen (Heidelberg: Carl Winter, 1979), 142–154.

90. Jonas saw in Heidegger a modern revival of the Gnostic alienation from the world and the direct expression of modern nihilism. See Hans Jonas, "Gnosticism and Modern Nihilism," *Social Research* 19 (1952): 452ff. The essay was incorporated into his *Gnostic Religion*, 2nd ed., "Epilogues: Gnosticism, Existentialism, and Nihilism." For exposition of Jonas's relationship to Heidegger's thought consult Vittorio Hösle,

"Hans Jonas's Position in the History of German Philosophy," in Tirosh-Samuelson and Wiese, eds., *The Legacy of Hans Jonas*, 19–38.

91. Jonas's dissertation, "Gnosis und spatantiker Geist," was completed in 1934 and appeared in English as *The Gnostic Religion: The Message of the Alien God and the Beginnings of Christianity* (Boston: Beacon, 1958).

92. There are various reasons for the relative obscurity of Jonas among Jews. First, Jonas had no official affiliation with a Jewish institution and no obvious constituency within the organized Jewish community in the United States. Second, Jonas's style was shaped by the European phenomenological tradition, which required considerable technical expertise in philosophy to which American Jews had difficulty relating. Third, Jonas focused on biology at a time when the Jewish community's intellectual interests were directed to political issues, especially the survival of the state of Israel. Therefore, even though Jonas articulated a post-Holocaust Jewish philosophy, most Jews remain unaware of it. For further discussion of the failure of Jonas to influence Jewish thought in the twentieth century see Hava Tirosh-Samuelson, "Understanding Jonas: An Interdisciplinary Project," in *The Legacy of Hans Jonas*, ed. Tirosh-Samuelson and Wiese, xxi–xlii, xxiii–xxvii.

93. See Hans Jonas, "Seventeenth Century and After: The Meaning of the Scientific and Technological Revolution," in his *Philosophical Essays: From Ancient Creed to Technological Man* (Englewood Cliffs, N.J.: Prentice-Hall, 1974), 45–80.

94. Jonas, *The Imperative of Responsibility: In Search of an Ethics for the Technological Age* (Chicago: University of Chicago Press, 1984), 1.

95. Ibid., 18–21.

96. Ibid., 18.

97. Jonas, *Philosophical Essays: From Ancient Creed to Technological Man*, 38.

98. Jonas, *Mortality and Morality*, 111.

99. Ibid., 101. For exposition of Jonas's understanding of responsibility consult Richard J. Rubinstein, "Rethinking Responsibility," *Hastings Center Report* 25:7, Special Issue (1995): 113–120.

100. Jonas, *Imperative of Responsibility*, 37.

101. Ibid., 42.

102. Ibid., 43.

103. Ibid., 43. Emphasis in the original.

104. Jonas, *Mortality and Morality*, 128.

105. For analysis of Jonas's use of kabbalistic ideas consult Ron Margolin, "Hans Jonas and Secular Religiosity," in Tirosh-Samuelson and Christian Wiese (eds.), *The Legacy of Hans Jonas*, 231–258.

106. For an excellent explanation of Jonas' analysis of metabolism consult Martin D. Yaffe, "Reason and Feeling in Hans Jonas's Existential Biology, Arne Naess's Deep Ecology and Spinoza's *Ethics*," in *The Legacy of Hans Jonas*, ed. Tirosh-Samuelson and Christian Wiese, 345–372, esp. 354–355; and Strachan Donnelley, "Hans Jonas and

Ernst Mayr: On Organic Life and Human Responsibility," in The Legacy of Hans Jonas, ed. Tirosh-Samuelson and Christian Wiese, 261–285.

107. Jonas, "The Burden and Blessing of Mortality," in *Mortality and Morality*, 87–98; originally published in *Hasting Center Report* 22:1 (1992).

108. Jonas, *Judaism and the Phenomenology of Life*, 36.

109. For a superb consideration of human uniqueness in light of contemporary science see J. Wentzel van Huyssteen, *Alone in the World: Human Uniqueness in Science and Theology: The Gifford Lectures* (Grand Rapids, Mich.: William B. Eerdmans Publishing Company, 2006).

110. The manifesto of the Transhumanist movement is to be found in James Hughes, *Citizen Cyborg: Why Democratic Societies Must Respond to the Redesigned Human of the Future* (Cambridge, Mass.: Westview, 2004).

SIX

The Bible and Biotechnology

LARRY ARNHART

We often assume that religion supports morality. Some people even think that if human beings did not believe that God commanded them to be moral, then morality would collapse. If God is dead, everything is permitted.[1]

I disagree, because I believe that human beings have a natural moral sense shaped by the evolutionary history of the human species. Charles Darwin was right in arguing that the moral sense is implanted in our human biological nature. Although religion can reinforce morality, our moral sense stands on its own natural ground independently of religious belief. In fact, morality is prior to religion in that we need to use our moral sense to judge the moral teachings of religion.

If the good is the desirable, I have argued, then morality is natural insofar as it satisfies the natural desires of our species. There are at least twenty natural desires that are manifested in diverse ways in all human societies throughout history, and these natural desires constitute the core of our natural moral sense. Human beings generally desire a complete life, parental care, sexual identity, sexual mating, familial bonding, friendship, social ranking, justice as reciprocity, political rule, war, health, beauty, wealth, speech, practical habituation, practical reasoning, practical arts, aesthetic pleasure, religious understanding, and intellectual understanding. Insofar as our moral sense is rooted in these natural desires, it does not *require* religious belief, although religious belief can *reinforce* the moral sense by confirming the lessons of nature. I have elaborated my arguments for these claims in my book *Darwinian Natural Right*.[2]

And yet, oddly enough, a few people have stubbornly refused to be persuaded by my absolutely demonstrative arguments. Some people have objected that my appeal to human nature as the ground of morality is threatened by the possibility that biotechnology could be used to alter, if not abolish, human nature. We are already hearing predictions about how biotechnology is leading us to "posthumanity." Can't we easily imagine how many people might be led by their natural desires to use biotechnology in ways that would eventually transcend the limits of human nature? For example, if we desire life, then why not use regenerative medicine to extend our lives indefinitely? And if we desire the health and happiness of our children, then why not genetically design those children to make them always healthy and happy? After all, if our human nature is only a product of a mindless process of Darwinian evolution that serves no moral purpose, then why shouldn't we use our technological power over nature to change our human nature to make it more to our liking? The only way to avoid this, some of my critics have insisted, is to invoke the religious belief that we were created by God in His image, and therefore to alter our given nature would transgress a divinely ordained order of existence.[3]

The dehumanizing consequences of liberating biotechnology from religious morality seem to be evident in a famous lecture by J. B. S. Haldane. One of the most prominent biologists of his day, Haldane gave a lecture in 1923 at Cambridge University with the title *Daedalus, or Science and the Future*.[4] In predicting the future applications of biology to human life, Haldane foresaw much of the subsequent history of biotechnology. He predicted that agricultural biotechnology would produce new forms of edible plants and animals so that there would be an abundance of food. Reproductive biotechnology would permit human eggs to be artificially fertilized, and then eventually human embryos could be grown in artificial wombs. This would allow sexual love to be separated from reproduction, which would liberate human sexual pleasure. This would also allow for artificial selection to improve the quality of human offspring with each new generation. New psychoactive drugs would permit human beings to artificially control their mental and spiritual faculties. Medical biotechnology would abolish most serious diseases and disabilities. Aging could be brought under human control to make it less abrupt and painful. For example, women could escape the effects of menopause by hormone replacement therapy so that they could prolong their youthful vigor and beauty.

Haldane saw that such technological changes in human nature would destroy traditional morality as supported by religion. Being a Marxist atheist, he insisted that "the scientific worker is not concerned with gods." He declared that the future technological conquest of nature would require a transformation of all values so that good would be turned into evil and evil into good. He explained: "We must learn not to take traditional morals too seriously. And it is just because even the least dogmatic of religions tends to associate itself with some kind of unalterable moral tradition, that there can be no truce between science and religion."[5]

For many people, Haldane's scientific future was a dreadful vision. His friend Aldous Huxley published *Brave New World* in 1932, which was a novel depicting a world similar to the future predicted by Haldane. But in Huxley's novel, this scientific utopia became a bleak, mechanized world in which people were dehumanized to the point of becoming soulless animals.[6] In 1947, C. S. Lewis published *The Abolition of Man*, in which he warned that a technological conquest of nature, like that described by Haldane, would produce a "world of post-humanity" that would mean the utter degradation of humanity.[7] More recently, critics of the biotechnological project for the conquest of nature such as Leon Kass and Francis Fukuyama have warned about the coming of a "brave new world" that would produce "the abolition of man."

These warnings have found their way into the political rhetoric of President George W. Bush, who gave a nationally televised speech in August of 2001 to warn that human embryonic stem cell research could bring a "brave new world" in which human beings would be created in test tubes.[8] In winning reelection in 2004, Bush used his opposition to embryonic stem cell research to attract the support of evangelical Protestants and conservative Catholics who fear biotechnology as an exercise in "playing God." Running through much of this alarm is the fear that biotechnology must be morally corrupting because modern science rejects the religious belief that, in the words of President Bush, "human life is a sacred gift from our Creator."[9]

Many of these critics of biotechnology appeal not just to religion in general, but to biblical religion in particular, as the only dependable source of moral limits on scientific technology. It is not surprising that Kass, the former chairman of the President's Council on Bioethics, has recently published a huge biblical commentary on the book of Genesis with the title *The Beginning of Wisdom*.[10] Kass's book shows how the Bible can be interpreted as providing religious reasons for limiting biotechnology. But it also shows the weaknesses in such an approach.

Any appeal to God as the supernatural source of morality creates more controversy than it resolves. There are three reasons for this. First, to rely on religious morality, we would have to agree on the existence and authority of God as demanding absolute obedience, but that provokes endless debate if we cannot agree on some natural ground of morality. Second, even if we could agree on God's existence and authority, we then would have to communicate with him to determine his will, but the language of divine communication—as in the biblical texts—is often so unclear that we cannot agree on its interpretation. Finally, even if we could agree on how a divine text like the Bible is to be interpreted, we would still have to judge its moral reliability, because sometimes the moral teaching of the text is clearly wrong. Ultimately, then, we have to pass the Bible—and other religious texts—through the filter of our natural moral sense.

Nature and Nature's God

Kass presents *The Beginning of Wisdom* as "a philosophic reading of the book of Genesis," with "philosophic" understood to mean "wisdom-seeking and wisdom-loving." In other words, he explains, he will read the Bible in "the same spirit in which I read Plato's *Republic* or Aristotle's *Nicomachean Ethics*—indeed, any great book—seeking wisdom regarding human life lived well in relation to the whole."[11] But then he immediately admits that this is a dubious undertaking because of the conflict between reason and revelation. "Religion and piety are one thing, philosophy and inquiry another. The latter seek wisdom looking to nature and relying on unaided human reason; the former offer wisdom based on divine revelation and relying on prophecy."[12] He also reports: "There are truths that I think I have discovered only with the Bible's help, and I know that my sympathies have shifted toward the biblical pole of the age-old tension between Athens and Jerusalem. I am no longer confident of the sufficiency of unaided human reason."[13]

This is confusing. How can he pursue a philosophic reading of the Bible relying on unaided human reasoning about nature, while losing confidence in the sufficiency of such reasoning? Kass acknowledges that a philosophical reader of the Bible would wonder how one can know that this is truly the revelation of God rather than a story made up by human beings. In response to such a demand for proof, the pious reader would insist that the true power of the Bible is revealed only to the faithful. To escape this impasse, Kass proposes to follow "a third alternative, an attitude between doubt, demanding proof in advance, and faith, comfortable that proof is unnecessary: the attitude of thoughtful engagement, of suspended disbelief, eager to learn."[14] So sometimes Kass stresses the tension between reason and revelation, which forces us to choose one side or the other. But at other times, he suggests overcoming this tension by finding some middle ground between the two.

That middle ground must be nature. By nature I mean the idea of the regular order of the observable world as knowable by human reason. As Kass indicates, the Hebrew Bible has no word for "nature," and this has been noted by some people who believe the very idea of nature is absent from the Bible. If everything is created by God, then we might think that everything exists not by any regular order but only by the contingent will of God. If there is no regular order in things, then philosophy or science as the inquiry into the causal regularity of the universe is futile. The only true wisdom would be unquestioning obedience to the arbitrary contingencies of God's will.

But it is possible that God has chosen to endow the world with a regular order that is comprehensible to the human mind, even though that order depends at every moment on His will as Ultimate Cause. If so, then the regular order of nature would be potentially comprehensible to the unassisted human reason of the philosopher or scientist, although access to the supernatural first cause of nature would require faithful submission to revelation as transcending the rational study of nature.

Kass is right to implicitly appeal to nature as a middle ground between reason and revelation. While indicating his nervousness about doing this, Kass explains that we need to see an idea of nature in the Bible "in order to bring our study of the biblical text into conversation with other wisdom-seeking activities."[15]

This idea of nature as common ground between scientific reason and biblical revelation underlies modern science. In the seventeenth century, Francis Bacon employed the theological metaphor of God's "two books" to sustain his vision for the new science of nature. To know God as "first cause" of all, Bacon argued, we must by faith look to the Bible as God's revelation, which surpasses human reason. But we can still use our human reason to inquire into the "second causes" by which God works in nature. To be wise, we need both religious faith in studying the Bible as "the book of God's word" and scientific inquiry in studying nature as "the book of God's work."[16] Charles Darwin quoted this passage from Bacon as the epigram for *The Origin of Species*.[17] Darwin then reaffirmed this thought at the end of the book by speaking of the natural evolution of life as displaying the "secondary causes" governed by "the laws impressed on matter by the Creator."[18]

If this is right, then those laws that the Creator has impressed on nature could include the moral laws that He has impressed on human nature. So while the revealed moral laws of the Bible would ultimately be supernatural and thus beyond unassisted human reason, there would also be natural moral laws that could be known by natural reason. Some Jews have seen a natural moral law in the Hebrew Bible as conveyed in the Noahide laws that are binding on all human beings.[19] Some Christians have seen a natural moral law in the New Testament as suggested by Paul's appeal to a natural law written in the hearts of all human beings.[20] Francis Hutcheson and other Scottish philosophers of the eighteenth century saw this natural moral law as rooted in the moral sense or moral sentiments of human nature.[21] Darwin extended this tradition of ethical naturalism by showing how the moral sense could have been implanted in human nature by natural selection in human evolutionary history.[22]

Although Kass sometimes seems to endorse the idea of the Noahide law as a natural law, he also suggests that the effectiveness of such a law depends on divine backing.[23] I agree that it helps the enforcement of a natural moral law to see it as divinely sanctioned. But what do we do when religious believers disagree about their religions and about the moral teachings of those religions? How do we settle moral disagreements between religious believers and those who are skeptical about religion? And how do we judge religiously based moral teachings that seem unreliable? In all such cases, we must appeal to some natural moral sense if we are to sustain a shared moral community.

Sometimes Kass points to a harmony between the Bible's account of human nature and the Darwinian view of human nature, which could support the idea of a natural moral sense. For example, he indicates that the Bible's presentation of the natural differences between men and women is confirmed by the modern Darwinian explanation of sex differences. That men are more promiscuous and aggressive in their sexual desire than women, that consequently male lust needs to be tamed by monogamous marriage so that it is directed to conjugal love and parental care, and that women typically invest more in producing and caring for children than do men—such traits of human biological nature are confirmed by both the Bible and Darwinian science.[24]

As part of the curse after the fall of Adam and Eve, God says to Eve: "I will terribly sharpen your birth pangs, in pain shall you bear children. And for your man shall be your longing, and he shall rule over you" (Genesis 3:16). Kass interprets this in the light of Darwinian evolutionary theory. Because of the evolutionary trend toward expansion of the human cranium, the newborn child's large head will cause the mother pain when passing through her relatively small birth canal. As compared with other species, the longer period of human gestation and of dependence of the young on parental care typically gives mothers a greater investment in their young than fathers, and this will incline mothers to focus intensely on securing the attachment of their husbands to themselves and their children. Kass speaks about "the female reproductive strategy, operative throughout the mammalian world: enlist all the help you can in support of your very few eggs and their living outcomes." He then explains: "How to gain the male's cooperation and permanent presence? How to domesticate him? A man who rules—or *appears* to rule—gets domestic authority in exchange for serving the needs of the woman and her children."[25]

Here and elsewhere in his commentary, Kass interprets the Bible as condemning erotic love that is not directed to producing children.[26] But other parts of the Bible would seem to work against this reading. The Song of Solomon is an intensely erotic love poem in which the lovers show no interest in procreation.

Kass reads the biblical teaching as a prediction of what will follow from the biological nature of men and women as sexual and reproductive animals. And thus he implicitly concedes that the moral authority of the Bible as divine revelation is insufficient, because faith in revelation must be confirmed by reasoning about nature.

A Cloudy Medium

In *The Federalist* (Number 37), James Madison observes that any written constitution must suffer from the inevitable obscurity and imprecision that come from using words to convey complex ideas. Even God cannot overcome these limitations in language. "When the Almighty himself condescends to address mankind in their own language, his meaning, luminous as it must be, is rendered dim and doubtful by the cloudy medium through which it is communicated."[27]

Some of the religious believers who fear biotechnology assume that the Bible provides a clear moral teaching that dictates severe limits on biotechnology. But when one looks for specific language in the Bible that might apply to biotechnology—or to science and technology generally—the "cloudy medium" of the language conveys only a "dim and doubtful" teaching.

The story of the Tower of Babel in the book of Genesis is often cited as condemning the technological mastery of nature as "playing God." And the biblical teaching that human beings are all equally created in God's image—and therefore that killing human life is a great crime—is interpreted as condemning the biotechnological manipulation of potential human life, as is done, for example, in embryonic stem cell research. But the actual language of the pertinent biblical texts is so vague that it supports conflicting interpretations.

In 1977, Jeremy Rifkin wrote a book attacking biotechnology with the title *Who Should Play God? The Artificial Creation of Life and What It Means for the Future of the Human Race*.[28] The title conveys the direction of his argument. The "creation of life" is proper only to God. For human beings to create life "artificially" is an arrogant transgression of God's law that will bring punishment upon the human race. Rifkin often uses the imagery of the Frankenstein story. Like Doctor Frankenstein, biotech scientists are trying to take God's place in creating life, and the result can only be the creation of monsters. So when people such as Rifkin use the phrase "playing God," they evoke a religious sense that nature is a sacred expression of God's will and therefore should not be changed by the human mastery of nature promised by modern science and technology. Rifkin has said that "the resacralization of nature stands before us as the great mission of the coming age."[29]

In contrast to Rifkin, Bacon defended the modern scientific mastery of nature as supported by biblical theology. Bacon argued that regarding nature as sacred was a pagan idea contrary to biblical religion. In pagan antiquity, the natural world was the sacred image of God, but the Bible teaches that God is the transcendent Creator of nature; and therefore, God's mysterious will is beyond nature. Although nature declares God's power and wisdom, it does not declare the will and true worship of God. Bacon believed that true religion as based on faith in biblical revelation must be separated from true philosophy as based on the rational study of nature's laws. And although the philosophic study of "second causes" in nature is no substitute for the knowledge of God as "first cause" revealed in the Bible to the faithful, we can use our natural knowledge of causes to master nature for human benefit, which conforms to our duty to love our neighbor. The Bible teaches us to employ our divine gift of reason not for aimless curiosity or overweening pride but "for the glory of the Creator and the relief of man's estate."[30]

The modern Jewish and Christian defenders of biotechnology agree with Bacon's biblical theology. Christian theologians such as Pope John Paul II, Philip Hefner, and Ted Peters read the Bible as teaching that human beings are "created cocreators." As "created," we are creatures and cannot create in the same way as God, who can create *ex nihilo*, "from nothing." But as "cocreators," we can contribute to changes in creation. Of course, we must do this as cautious and respectful stewards of God's creation, but the Bible teaches us not to worship nature as sacred and thus inviolable.[31]

Which side in this debate is correct? Does the Bible condemn or support the scientific mastery of nature through technology, including biotechnology? Kass and others would say that the story of the Tower of Babel is the primary biblical text on this question, and it clearly supports Rifkin's position rather than Bacon's. But I am not so sure.

Here is the whole text in Genesis 11:1–9 as translated by Robert Alter.

> And all the earth was one language, one set of words. And it happened as they journeyed from the east that they found a valley in the land of Shinar and settled there. And they said to each other, "Come, let us bake bricks and burn them hard." And the brick served them as stone, and bitumen served them as mortar. And they said, "Come, let us build a city and a tower with its top in the heavens, that we may make us a name, lest we be scattered over all the earth." And the Lord came down to see the city and the tower that the human creatures had built. And the Lord said, "As one people with one language for all, if this is what they have begun to do, now nothing they plot to do will elude them. Come, let us go down and baffle their language there so they will not understand each other's language." And the Lord scattered them from there over all the earth, and they left off building the city. Therefore, it is called Babel, for there the Lord made the language of all the earth babble. And from there the Lord scattered them over all the earth.[32]

According to Kass, this story shows the Bible condemning "the universal, technological, secular city," in which human beings are "trying to play God." It is a warning against all civilization as founded on agriculture and urban life, in which human beings are "taking the place of God."[33] Moreover, Kass claims that "the project of Babel has been making a comeback" in the modern Baconian project for using science and technology to master nature for the relief of the human estate. In fact, all of modern life seems dedicated to this new "project of Babel." Kass observes:

> Whether we think of the heavenly city of the philosophes or the posthistorical age toward which Marxism points, or, more concretely, the imposing

> building of the United Nations that stands today in America's first city; whether we look at the World Wide Web and its WordPerfect, or the globalized economy, or the biomedical project to re-create human nature without its imperfections; whether we confront the spread of the postmodern claim that all truth is of human creation—we see everywhere evidence of the revived Babylonian vision.[34]

We must wonder whether Kass has really discovered this as the clear teaching of the biblical text, or whether he has brought to the text his own scorn for modern life. Kass actually admits that the story of the Tower of Babel has "multiple ambiguities" and that some of his reading of the story "does not square neatly with the text."[35]

It is worth noting that Jean-Jacques Rousseau is the one philosopher whom Kass cites most often.[36] One might wonder whether Kass's denunciation of modern civilization shows not so much the influence of the Bible as the influence of Rousseau's romantic rejection of the arts and sciences as morally corrupting.

Most readers could probably agree with Robert Alter's observation about the Tower of Babel that "the polemic thrust of the story is against urbanism and the overweening confidence of humanity in the feats of technology."[37] But it is not so clear that this story shows the Bible to be condemning all civilization—and particularly all modern civilization—as evil. After all, if this were true, the Bible would be condemning itself, because the Bible as a written text passed down through the centuries presupposes the invention of writing and—in the modern world—the technology of printing, as well as the complex social and religious institutions of civilization that create and preserve written texts.

The story of the Tower of Babel contributes to a biblical theme of scorn for cities and urban civilization generally. The first murderer in the Bible—Cain—is also the first founder of a city (Genesis 4:17). Babel (or Babylon) was founded by Nimrod, a descendant of Noah's son Ham (Genesis 10:8–12). The descendants of Ham seem generally to be city-builders. By contrast, the descendants of Noah's son Shem are generally rural or mountain people, and this includes Abraham, who will be called by God as the patriarch of the Chosen People. The political and intellectual development in cities seems to cultivate a proud artfulness in city people contrary to the humble simplicity of rural people that is more conducive to pious obedience to God.

Babylon and other cities in Mesopotamia were the first human settlements to show the move from foraging to farming, as early as eleven thousand years ago. As Kass suggests, the Bible recognizes agriculture as the crucial innovation that made civilization possible. "Farming requires intellectual sophistication and psychic discipline." And agriculture "represents a giant step toward human self-sufficiency."[38]

If we define biotechnology broadly as the technical manipulation of organisms to produce products and services to satisfy human desires, then we can recognize the agricultural manipulation of plants and animals as the beginning of biotechnology. For most of their evolutionary history, human beings have lived as foragers who hunted wild animals and gathered wild plants, which required an unsettled existence that could not sustain a civilized, urban life. But with the invention of agriculture in ancient Mesopotamia, human beings produced food by herding domesticated animals and cultivating domesticated plants. As a result, these farmers and herders were breeding domesticated plants and animals to make them more suitable for human use, and in doing so, they were artificially selecting for genetic modifications in these domesticated species. Even today, all of human civilization depends on this project in agricultural biotechnology for the abundant production of food.[39]

Many of the innovations today in biotechnology continue to come from agricultural biotechnology. For example, in vitro fertilization and cloning (including the cloning of Dolly the sheep) were first done by agricultural researchers to develop more productive farm animals. In England, John Hammond, an agricultural scientist, developed techniques for improving livestock through in vitro fertilization as early as the 1920s. He was opposed by some religious leaders in the Church of England who condemned the artificial insemination of farm animals as an immoral technology because it separated sex from reproduction.[40]

It might seem, then, that Kass is right in seeing the biblical condemnation of cities and agriculture in the story of the Tower of Babel as a general condemnation of science and technology. But this is not so clear when one considers the rest of the Bible. Isaiah celebrates Jerusalem as God's city for his Chosen People, in contrast to the evil city of Babylon (Isaiah 24–26, 60–62). So it seems that not all cities are evil. Kass's response is that God's selection of Jerusalem as his city is only a concession to human weakness, but Kass cites no biblical text to support this assertion.[41]

The Bible praises Solomon for his God-given wisdom, which includes a natural knowledge of plants and animals and a technological knowledge that allows him to build his great palace and the Temple of God. The Bible recounts in intricate detail the craftsmanship that Solomon employed during the twenty years of his building projects (I Kings 5–9). For Bacon, Solomon was a biblical model for knowledge and wisdom that embraced natural philosophy and the practical arts.[42]

Many of the founders of modern science—such as Robert Boyle and Isaac Newton—were careful readers of the Bible, and they believed that in discovering the causal laws of nature, they were exploring the mind of God. Some historians of science have confirmed Alfred North Whitehead's observation that science as it arose in the Western world was sustained by the biblical teaching that God created the universe as a fully rational order.[43] If Kass is right, this is incorrect because the Bible actually condemns scientific knowledge as a sinful manifestation of human pride. So, again, it seems that the language of the Bible is a "cloudy medium" that is open to conflicting interpretations.

The same problem of vagueness arises when one tries to discern the Bible's teaching on specific moral issues in biotechnology. Consider, for example, the moral debate over research with human embryonic stem cells. Since embryonic stem cells have the potential to develop into any of the specialized cells of the human body, it is possible that stem cells could some day be used to replace or repair damaged, defective, or aging cells everywhere in the body. Consequently, stem cell therapy could relieve human suffering, preserve life, and perhaps even extend the human life span.

The moral problem is that extracting stem cells from human embryos destroys the embryos. Such embryos are tiny blastocysts with a hundred or more cells that develop twelve to fourteen days after the fertilization of the egg. Some people say that since these blastocysts are far from being fully human, destroying them to promote potentially lifesaving medical therapies is morally acceptable. Some Jewish and Christian theologians have taken this position as compatible with the Bible. But other people insist that destroying a human embryo violates the biblical law against murdering human beings.

So what does the Bible say about this? The quick answer to this question would be "Nothing." After all, the Bible says nothing about embryonic stem cells or the possibility of stem cell research. If the Bible is God's revelation of His will, then we must say that God chose not to declare any moral law governing embryonic stem cell research. If there is any implied biblical law on this subject, we must try to interpret passages in the Bible that do not speak explicitly about this, and consequently we will be left in uncertainty.

The religious debate over the moral status of embryonic stem cells is intertwined with the debate over abortion and the question of whether human life begins at conception. In the 1995 encyclical *Evangelium Vitae* on abortion and euthanasia, Pope John Paul II declares that human life begins at conception. And yet he admits that there is no clear biblical statement of this point. "There are no direct and explicit calls to protect human life at its very beginning, specifically life not yet born." Moreover, he admits that "the texts of Sacred Scripture never address the question of deliberate abortion and so do not directly and specifically condemn it."[44] Consequently, the pope and others who seek a moral teaching on the beginning of human life from the Bible are forced to draw indirect inferences from texts that do not directly state a conclusion on this issue.

Isaiah says that God "called me when I was in the womb, before my birth he had pronounced my name," and further, that God "formed me in the womb to be his servant, to bring Jacob back to him and to reunite Israel to him" (Isaiah 49:1, 5). Although this suggests a divine predestination sometime before birth, it is not clear that this means that full human life begins immediately upon the fertilization of the mother's egg. A similar predestined calling is declared by God to Jeremiah: "Before I formed you in the womb, I knew you; before you came to birth I consecrated you; I appointed you as prophet to the nations" (Jeremiah 1:4–5).

Job says to God: "Did you not pour me out like milk, and then let me thicken like curds, clothe me with skin and flesh, and weave me of bone and sinew?" (Job 10:10–11). But, again, it is not clear that this means that full human life begins at conception.

In the account of the creation of humanity, it is said "the Lord God fashioned the human, humus from the soil, and blew into his nostrils the breath of life, and the human became a living creature" (Genesis 2:7). Some biblical believers have interpreted this distinction between earthy material and breath of life to suggest that full humanity begins in the womb only with the formation of lungs and nostrils. But, again, it is not clear that this is the intended meaning of the biblical text.

As part of the Mosaic law, it is stated that "should men brawl and collide with a pregnant woman and her fetus come out but there be no other mishap, he shall surely be punished according to what the woman's husband imposes upon him, he shall pay by the reckoning. And if there is a mishap, you shall pay a life for a life, an eye for an eye, a tooth for a tooth, a hand for a hand, a foot for a foot, a burn for a burn, a wound for a wound, a bruise for a bruise" (Exodus 21:22–26). Here it seems that the loss of a pregnancy is not treated as the loss of a human life but as a loss to the woman's husband, which requires a monetary penalty rather than capital punishment, and thus by implication, the killing of a fetus is not murder, because a fetus is not fully human. But, again, the text is unclear.

In the moral debate over human embryonic stem cell research, many Jews and liberal Protestants have favored such research because of the possible medical benefits, and because the human blastocyst is not fully human. On the other side, many conservative Catholics and evangelical Protestants have opposed such research because they think that killing a human blastocyst is murder.

Representing the Jewish position, particularly within the Conservative movement of Judaism, Rabbi Elliot Dorff cites the biblical statement that human beings were put in the Garden of Eden "to till it and watch it" (Genesis 2:15). He reads this as suggesting that we are supposed to both work on the world and preserve it, which requires a balance between mastery and humility. This includes the duty to preserve the life and health of our fellow human beings. In support of this duty to heal, Dorff cites the biblical injunctions "You shall not stand idly by the blood of your fellow man," and "You shall love your fellow man as yourself" (Leviticus 19:16, 18). Since he believes that the Bible does not recognize the human embryo as having the moral status of a full human being, Dorff concludes that our moral duty to heal makes it a duty to use embryonic stem cells to promote life and health.[45] Kass has expressed his frustration that so many of his fellow Jews do not share his fear of biotechnology.[46]

Lutheran theologian Ted Peters agrees with Dorff that using embryonic stem cells for medical purposes fulfills the biblical duty to love our neighbor. Peters argues that the early human embryo has many life-giving potentials. If it is flushed from the uterus, as most fertilized eggs are, then it creates no life. If it remains in the uterus and divides, it produces twins. If it divides a second time, it produces quadruplets. If it does not divide, it can develop into one distinct person. If it is outside a uterus, and its stem cells are extracted, these stem cells have the potential to contribute to human healing. Thus, Peters argues, a fertilized human egg has many potential lines of life-giving development, and producing stem cells is one of them.[47]

So here we have an intense moral debate over biotechnology in which all sides claim to have biblical authority for their position. This happens because the Bible is sometimes unclear about some moral issues. In such cases, we must exercise our moral judgment independently of the Bible.

Biblical Immorality

Not only is the Bible's moral teaching sometimes unclear, but it is also sometimes unreliable. When the Bible's teaching is immoral, we must correct it by appealing to our natural moral sense.

Rabbi Dorff observes that "some passages in the Bible are morally ambiguous at best and downright immoral at worst—texts such as God's commands to bind and presumably murder Isaac, to kill all of the Canaanites as part of the conquest of the Holy Land, and to enslave non-Jews." In such cases, he contends, we must use our moral judgment to correct the Bible.[48]

Consider his three examples of biblical immorality. The first is the story of God commanding Abraham to kill his son Isaac as a sacrifice. Earlier God has made a covenant with Abraham, promising that through his son Isaac, Abraham will be father to a multitude of nations and kings that will live in the land of Canaan forever as God's chosen people (Genesis 17:1–22). Later, it is said that "God tested Abraham" by commanding him to take his son to a mountain and offer him up as a burnt offering. Abraham obeys. He travels for three days to the mountain. There he binds Isaac on a pile of wood on an altar. And finally he raises a butcher knife to slaughter his son. But then he hears an angel who tells him "Do not reach out your hand against the lad, and do nothing to him, for now I know that you fear God, and you have not held back your son, your only one, from Me." Abraham then sees a ram that he sacrifices in place of his son. Then, the angel speaks again: "Because you have done this thing and have not held back your son, your only one, I will greatly bless you and will greatly multiply your seed, as the stars in the heavens and as the sand on the shore of the sea, and your seed shall take hold of its enemies' gate. And all the nations of the earth will be blessed through your seed because you have listened to my voice" (Genesis 22:12–17).

Readers of this biblical story have found it deeply disturbing. First, Abraham is willing to obey a command from God that contradicts God's earlier covenant with Abraham to give him perpetual descendants through Isaac. And even more troubling is that God orders a father to murder his innocent child, the father obeys, and then God praises him for being willing to commit such a murder.

Many readers have tried to make this story morally comprehensible.[49] Kass suggests that this was a proper test of Abraham's fear of God, and "it is the fear of the Lord that is the beginning of wisdom."[50] But Kass admits that he is not confident about his interpretation of this "shocking story." He even suggests that Isaac had "good reason" not to understand or approve of what his father had done.[51]

Søren Kierkegaard in *Fear and Trembling* interprets Abraham's binding of Isaac as showing that true faith in God requires a "teleological suspension of the ethical." The faithful individual must obey any command from God, even when that command violates the universal principles of morality. To judge the morality of God's commands shows a lack of faith because it denies the arbitrary omnipotence of God's will. If we truly love God, we will obey his every command without question, even when He commands us to murder an innocent child. For a loving father to sacrifice his son for a high ethical principle would be tragic and yet glorious and comprehensible, because we could say that a lower moral duty yielded to a higher moral duty. And yet while this would make the father a "tragic hero," this would not make him a "knight of faith." Abraham's act was utterly different. He was willing to kill his son not for the sake of a higher moral duty but for the sake of obeying God as the omnipotent will that can suspend all principles of moral duty. The conduct of the "tragic hero" is humanly comprehensible as showing the conflict of moral duties. But the conduct of Abraham as the "knight of faith" is not morally comprehensible at all. A loving father was ready to murder his son simply because he heard God's voice commanding him to do this. "Humanly speaking," Kierkegaard says, "he is crazy and cannot make himself intelligible to anyone."[52]

Kierkegaard insists that this "suspension of the ethical" in unquestioning obedience to God's commands is declared not only in the Old Testament but in the New Testament as well. Jesus taught: "If any man cometh unto me and hateth not his own father and mother and wife and children and brethren and sisters, yea, and his own life also, he cannot be my disciple" (Luke 14:26).[53] Thus does Jesus affirm the same terrifying teaching as in the story of Abraham's binding of Isaac.

The incomprehensibility of Kierkegaard's "knight of faith" should make us wonder whether this is anything more than an intellectual fantasy that goes beyond the human experience of faith. Kierkegaard almost concedes this: "I candidly admit that in my practice, I have not found any reliable example of the knight of faith, though I would not therefore deny that every second man may be such an example."[54]

What Kierkegaard saw as the purest expression of faith, Immanuel Kant saw as the clearest example of religious delusion. Since God is just, He cannot command anything that would be unjust. Therefore, any man who would hear a voice from heaven commanding him to murder his child should know that this is an illusion and not a true revelation.[55] The story of Abraham shows the immorality of religious fanaticism unconstrained by moral law. Kant would agree with Kierkegaard on one point: Abraham really was crazy.

Ronald Green is a scholar of religion who has argued that all religions are rooted in a universal structure of moral reasoning that is common to all human beings. And yet he admits "religious traditions also sometimes incorporate immoral norms."[56] Conceding that the biblical story of Abraham's binding of Isaac provides "the hardest case of all" for his claim that all religion supports morality, Green surveys the long history of Jewish and Christian commentators finding ways to interpret the story so that it is morally comprehensible. For example, some Jewish commentators argued that Isaac consented to the sacrifice, and therefore that this was a case not of murder but of martyrdom. And many Christian commentators followed the lead of Paul who declared that Abraham had faith in God's power to resurrect Isaac from death, and therefore killing his son would not be murder (Hebrews 11:17–19). Thus, the commentators added something to the story to justify what otherwise would be a disturbing case of God ordering a father to murder his son as a religious sacrifice.

Thomas Aquinas strained to make the binding of Isaac consistent with the Mosaic commandment against murder. He reasoned that since all human beings deserve death from God because of the original sin of Adam, any human being can be authorized by God to carry out that original death sentence. Thus, when Abraham obeyed the commandment to kill Isaac, he did not consent to murder, which would be killing the innocent, because Isaac—like all human beings—was guilty of original sin.[57] This is a strange line of reasoning, because it uses the doctrine of original sin to suspend the prohibition against murder.

As Green says, these interpretations illustrate "the ways rational and moral considerations mold a tradition ostensibly based on revelation and the divine command."[58] Here the phrase "ostensibly based" implies that while Kierkegaard was right about the "suspension of the ethical" in this biblical story of "revelation and divine command," the "tradition" of religious commentary on this story has corrected the biblical text by using "rational and moral considerations" to cover up its immoral teaching.

After Abraham has passed his test of faith, the angel restates God's earlier promise that Abraham's descendants will become powerful, while adding a promise that they will have military victories over their opponents: "your seed shall take hold of its enemies' gate" (Genesis 22:17). To fulfill this promise, Moses later commanded a military policy of genocidal holy war against the people living in Palestine. Everyone must be killed—men, women, and children. "You shall let no breathing creature live" (Deuteronomy 20:16). In his comments on this passage, Robert Alter worries about its immorality. "It is hard to find any mitigation for the ferocity of this injunction to total destruction. The rabbis reinterpreted it, seeking to show that it was almost never strictly applicable."[59]

When the Israelites defeated the Mideanites, the Israelite soldiers executed all of the adult males. But Moses was angry that they had been too merciful. He ordered: "And now, kill every male among the little ones, and every woman who has known a man in lying with a male. And all the little ones of the women who have not known lying with a male, let live" (Numbers 31:17–18). Alter observes: "Moses's command—one should note that it is Moses's, not God's—to perpetrate this general massacre, excluding only virgin females, is bloodcurdling, and the attempts of the interpreters, traditional and modern, to 'explain' it invariably lead to strained apologetics."[60]

Here we see one of many examples in the Bible of the Machiavellian use of cruelty.[61] Alter suggests that the biblical story of David is "the first full-length portrait of a Machiavellian prince in Western literature."[62] Actually, Machiavelli admired both Moses and David.[63] As with the binding of Isaac, religious believers have had to "reinterpret" the biblical teaching of holy war and Machiavellian politics to correct its immorality.

Sometimes, the Israelites enslaved their enemies rather than kill them, and these people could be permanently enslaved (Leviticus 25:35–55; Joshua 9:19–27). Both the Old and the New Testament support slavery. Paul admonished slaves that they had a duty to obey their masters (Ephesians 6:5–9). An important element of the proslavery arguments in the American South before the Civil War was that slavery was based on the laws of God as conveyed in the Bible.[64] The modern assumption that the Bible teaches the universal moral equality of all human beings has arisen largely from the strained interpretations of the Bible by libertarian political philosophers like John Locke. In arguing that biblical religion supports the natural liberty and equality of all human beings, Locke was careful in his biblical citations to draw his reader's attention away from the biblical support for slavery.[65]

In *The Abolition of Man*, C. S. Lewis argues that all morality rests upon some universal principles of "traditional morality" or "natural law" that he calls the *Tao*. In the appendix to the book, he offers some illustrations of these principles with quotations from religious texts, philosophic works, and literary works throughout history. His biblical quotations are highly selective. For example, to illustrate "the law of general beneficence," he quotes Exodus 20:13: "Do not murder."[66] But he does not quote Deuteronomy 20:16: "You shall let no breathing creature live." His selectivity shows him using his moral judgment to pick out the moral teachings in the Bible while passing over in silence the immoral teachings. Although Lewis was a devout Christian believer in the Bible, his moral sense did not depend upon the Bible.

For the Bible to be an authoritative, clear, and reliable guide to moral judgment, we must read the Bible in the light of our natural moral sense. We can thus use biblical religion to reinforce our moral sense, even though that moral sense stands on its own independently of the Bible, which allows us to correct the mistakes in the Bible's moral teaching. This moral sense is natural to us because it is rooted in our human biological nature as shaped by the evolutionary history of our species.

Actually, the moral teaching of the Bible generally agrees with this natural moral sense. I have pointed to some examples of biblical teachings that we can recognize as immoral. But what is remarkable is that most of the general principles of the Mosaic law conform to natural morality. Aquinas could argue that the moral teaching of the Mosaic law—as distinguished from the ceremonial and judicial precepts that were peculiar to the Israelites—were part of the natural law binding on all human beings by virtue of their human nature.[67] Moses himself suggested as much in the book of Deuteronomy, which is a series of valedictory speeches that Moses delivered to the Israelites shortly before his death. He declared that the practical wisdom of his laws would become evident to all human beings, even those who did not worship the true God (Deuteronomy 4:6). All human beings in principle could recognize the prudential goodness of those moral laws that were adapted to the natural human condition. To justify his laws, Moses repeatedly insisted that if the Israelites obeyed his laws, they and their children would survive and prosper in their new land. He made no claims about immortality of the soul or about rewards and punishments in an afterlife. Instead, like Charles Darwin, he assumed that the purpose of morality was to secure the earthly survival and prosperity of oneself and one's progeny—survival and reproduction. Promoting the survival and reproduction of the Jews required social norms so that individuals would cooperate within their group to compete with other groups (Deuteronomy 4:1–6, 4:40, 5:29, 5:33, 6:1–3, 6:24, 8:1, 11:8–9, 11:21, 22:7, 23:9–14, 25:15, 30:15–20). Jesus reiterated this in teaching that to secure earthly life, human beings would need to obey the Second Table of the Ten Commandments—the moral rules for no murder, no adultery, no stealing, no giving false testimony, honoring parents, and generally loving one's neighbor as oneself (Matthew 19:16–19).

As a product of natural human experience, not only Judaism but all religions serve the natural desires of human beings in diverse social and physical environments, and consequently we can explain religion as an adaptation of human ecology. From a Darwinian point of view, religion has functional social utility because it makes religious believers more cooperative within their religious groups so as to promote survival and reproduction.[68] Religious morality is rooted in human nature.

But what happens if biotechnology gives us the power to change, or even abolish, our human nature? If human nature itself becomes mere matter to be mastered by our technological power to create what C. S. Lewis called "the world of post-humanity," does that mean that we will no longer have human nature as a source for a natural moral sense? From Lewis and Aldous Huxley to Leon Kass and Francis Fukuyama, the dread of modern biotechnology has arisen from this fear that "the abolition of man" will obliterate the natural ground of moral experience, because human nature itself will become a human artifact.

Some people would say that the only way to avoid such a dreadful outcome is to learn from the Bible that a biotechnological mastery of human nature would be a new Tower of Babel that would bring a disastrous punishment for our excessive human pride. But, as I have argued, it is not clear that the Bible has enough authority, clarity, and reliability in its moral teaching to resolve our debates over biotechnology.

If biblical religion is insufficient to solve our moral problems with biotechnology, then we must turn to our natural moral experience. And our first step in applying our natural moral sense to biotechnology is to see that the power of biotechnology for changing human nature has been exaggerated. The most fervent advocates of biotechnology welcome any prospect of using it to transform our nature to make us superhuman. The most fervent critics of biotechnology warn us that its power for transforming our nature will seduce us into a Faustian bargain that will dehumanize us. Both sides agree that biotechnology is leading us to a "posthuman future." But that assumption is false. It ignores how Darwinian evolution has shaped the adaptive complexity of our human nature—our bodies, our brains, and our desires—in ways that resist technological manipulation. A Darwinian view of human nature reveals the limits of biotechnology so that we can reject both the redemptive hopes of its advocates and the apocalyptic fears of its critics.

Natural Means

If we keep in mind the adaptive complexity of human nature, we can foresee that biotechnology will be naturally limited both in its technical means and in its moral ends. It will be limited in its technical means, because complex behavioral traits are rooted in the intricate interplay of many genes interacting with developmental contingencies and unique life histories to form brains that respond flexibly to changing circumstances. Consequently, precise technological manipulation of human nature to enhance desirable traits while avoiding undesirable side effects will be very difficult if not impossible. Biotechnology will also be limited in its moral ends, because the motivation for biotechnological manipulations will come from the same natural desires that have always characterized human nature.[69]

The strongest proponents of biotechnology are led by libertarians such as Lee Silver, Gregory Stock, and Ronald Bailey. The strongest opponents of biotechnology are led by neoconservatives such as Leon Kass, Francis Fukuyama, and William Kristol, and by environmentalists such as Bill McKibben and Jeremy Rifkin. They disagree in that the proponents are hopeful about the future to be created by biotechnology, while the opponents are fearful. But both sides agree that biotechnology is giving human beings the technical power to transform human nature and move toward "posthumanity." I disagree because I think both sides in this debate employ a rhetoric of exaggeration that ignores the technical limits on biotechnology.

As suggested by the title of his book *Remaking Eden*, Lee Silver foresees that biotechnology will soon give us the godlike power to recreate ourselves into whatever form we might want.[70] Silver begins each part of his book with a quotation from the Bible. But he does this ironically to suggest that the power to create life can no longer be limited by biblical religion.[71] He indicates that the ultimate objection to biotechnology belongs to the "realm of spirituality," because it's the objection that biotechnology is "treading in God's domain." He insists, however, that once "man's domain" is extended into control of the human genome, then "God's domain vanishes into . . . nothingness."[72] Silver agrees with Kass that the biblical story of the Tower of Babel is a warning against human beings using their technological knowledge to gain complete power over nature.[73] But unlike Kass, Silver believes that biotechnological mastery of nature is too attractive to be stopped by biblical stories.

Once we have crossed the line into gaining human power over the genetic basis of life, Silver claims, the use of that power to satisfy human desires is inevitable. The only moral limit that remains is the libertarian principle that we must respect individual freedom of choice, so that we allow people to use reproductive technology in any way they choose, as long as it does not directly harm anyone else.[74]

This will permit parents to use genetic engineering to create "designer babies" that will have the traits of body and mind desired by the parents. But Silver is never very clear about how exactly this is going to happen. He observes: "Some of the ideas proposed here may ultimately be technically impossible or exceedingly difficult to implement. On the other hand, there are sure to be technological breakthroughs that no one can imagine now."[75] So it's impossible now, but it might become possible in the future with unpredictable "technological breakthroughs." This kind of rhetorical move is common among the writers on biotechnology. It allows them to spin out imaginative scenarios based on speculative assumptions about the future. For example, the last third of Silver's book becomes an exercise in science fiction as he imagines evermore bizarre possibilities for the future.

Gregory Stock agrees with Silver in defending a libertarian stance on human reproductive biotechnology. He also exaggerates in the same ways as Silver. In his book *Redesigning Humans*, Stock declares that "the arrival of safe, reliable germline technology will signal the beginning of human self-design." He admits, however, that "our biology might prove too complex to rework." He concedes that "no present genetic intervention is worth doing in a healthy individual, and no present technology is capable of effecting an intervention safely anyway." He acknowledges that many biologists believe that the genetic propensities underlying complex behavioral traits such as personality and intelligence are so intricate that we could never intervene to change these mechanisms without producing undesirable side effects. He also recognizes that these genetic propensities always interact in unpredictable ways with chance events and life history to produce unique individuals in ways that cannot be controlled by genetic technology. "Even for highly heritable traits," he observes, "it will be uncertain what a child's unique amalgam of potential and experience will bring. A vision of parents sitting before a catalog and picking out the personality of their future 'designer child' is false."[76]

But then just when it seems that he has given up on the idea of "human self-design" through genetic technology, Stock suggests, "technological barriers soon may fall." We might someday develop an artificial human chromosome, and we might find a way to use it to safely change the complex behavior traits of our children through genetic engineering. In describing these technological novelties of the future, he uses words such as "may," "might," "probably," and "ideally" in almost every paragraph of his writing. He writes: "At this time, human germline manipulation is not feasible or safe. A decade from now, it still won't be. Two or three decades hence, however, the story may be different."[77] So in thirty years, we might be able to do what today is impossible. Well, maybe. Or, maybe not. As with Silver, it's hard to know how to respond to Stock's speculative scenarios for the future, except to identify them as nothing more than speculative scenarios.

Bill McKibben would seem to be radically opposed to the position of Silver and Stock. In his recent book *Enough: Staying Human in an Engineered Age*, McKibben argues that we need to limit our technological power over human life by deciding that we don't need more such power because we already have "enough." If we don't say "enough," then our growing biotechnological power will soon destroy our human identity, we will become like robots, and life will be meaningless.[78] He thinks that "this idea of restraint comes in large measure from our religious heritage," which should teach us that "meaning counts, more than ability or achievement or accumulation."[79]

And yet the urgency of McKibben's argument depends on his agreeing with Silver and Stock that we are headed toward the abolition of human nature through biotechnology. In much of his book, he simply paraphrases or quotes from Silver, Stock, and others who think technology is moving us to posthumanity. He then concludes: "The technoprophets have made a persuasive case that we will soon be able to leave humanness behind." An example of his naïve acceptance of even the wildest speculations is his report that scientists "are already hot on the trail of a human 'happiness gene,'" and therefore, "it's not particularly far out to imagine genetic engineering designed to make our children happier."[80]

Like McKibben, neoconservatives like Kass and Francis Fukuyama accept the biotechnological prophecies of people such as Silver and Stock, because they agree that, as Kass says, biotechnology is "a runaway train now headed for a post-human world."[81] And yet Fukuyama admits—in his book *Our Posthuman Future*—that "we do not today have the ability to modify human nature in any significant way, and it may turn out that the human race will never achieve this ability." He then adds, however, that genetic engineering to change human nature might become possible in a hundred years.[82] Well, maybe. Or, maybe not.

The people on both sides of this debate agree that modern biotechnology fulfills Francis Bacon's original project for mastering nature through technological science. But they fail to stress Bacon's recognition that the human conquest of nature will always be limited in its technical means by the potentialities of nature itself. Bacon observed that "nature to be commanded must be obeyed," because "all that man can do is to put together or put asunder natural bodies," and then "the rest is done by nature working within."[83] In 1985, Kass in his book *Toward a More Natural Science* used these words of Bacon—without identifying them as Bacon's—in explaining how the power of biotechnology is limited by the potentialities inherent in nature itself.[84] But then, in his later writings, he has depicted biotechnology as unlimited in its technical power.

Throughout the history of biotechnology—from the ancient Mesopotamian breeders of plants and animals, to Louis Pasteur's use of microorganisms for fermentation and vaccination, to Herbert Boyer and Stanley Cohen's techniques for gene splicing—people have employed nature's properties for the satisfaction of human desires.[85] Boyer and Cohen did not create restriction enzymes and bacterial plasmids. They discovered them as parts of living nature. They then used those natural processes to bring about outcomes—such as the bacterial production of human insulin for people with diabetes—that would benefit human beings. Baconian biotechnology is thus naturally limited in its technical means because it is constrained by the potentialities of nature.

Both proponents and opponents of biotechnology ignore these natural limits. For example, in predicting a future of "designer children," McKibben ignores the adaptive complexity of mental traits that arise from many interacting and unpredictable causes in ways that are not amenable to precise genetic manipulation. McKibben predicts that soon parents will be able to increase the innate intelligence of their children by selecting those genes that enhance intelligence. He cites the work of psychologist Robert Plomin who announced in a 1998 article that differences in a gene on chromosome 6 could account for 2 percent of the difference between a group of children with high IQ scores and another group with lower scores.[86] McKibben fails, however, to tell his readers that Plomin's finding has never been replicated by any other researchers. Furthermore, in the fall of 2002, Plomin retracted his 1998 report, because he had failed to replicate it himself.[87] Recently, Plomin has admitted that in searching for genes that influence intelligence, "the track record for replicating gene associations is not good." He has also conceded that since intelligence is controlled by so many genes, with each gene exerting only a small effect, it might be impossible to identify exactly the genetic basis of intelligence.[88]

When we speak of the "genetic basis" of intelligence or some other complex human trait, we should remember that we are talking about genetic *propensities* or *potentialities*, not genetic *determinism*. Although genes are a *necessary* cause of these traits, the genes are not the *sufficient* cause. Even if we could explain exactly the multiple genetic causes of intelligence, we would still have to explain how those genes influence neural activity and how genetic propensities and neural activity interact with environmental contingencies in the unique life histories of particular human beings. And all of this would presuppose that we could agree on how to define and measure "intelligence," even though both scientific research and commonsense experience suggest that there are different kinds of intelligence—for example, analytic intelligence, verbal intelligence, practical intelligence, musical intelligence, and kinesthetic intelligence.[89] We should also recognize that since our intellectual activity arises from a complex interaction of reason and emotion, we cannot explain intelligence without also explaining emotion.[90]

The Bible confirms the importance of intelligence as the defining human trait. As created in the "image of God," human beings are godlike. And what distinguishes God in the Bible is his capacity for intelligence and intelligent choice. As Kass says, "God exercises speech and reason, freedom in doing and making, and the powers of contemplation, judgment, and care."[91] Human beings are elevated above the other creatures to the degree that human beings share in these godlike traits of intelligence, and yet they can never attain the level of God's mind.

Could we use biotechnology to redesign ourselves so that we might rival God in intelligence? Surely biotechnology can give us some control over the biological bases of intelligence—the genetic, physiological, and neurological mechanisms that support human reasoning—at least to the point of preventing or treating mental retardation caused by biological disorders. But the complex nature of intelligence makes it unlikely that our biotechnological control could ever be so precise and complete that we could radically improve or transform our normal human range of intelligence. Contrary to the predictions of McKibben and others, it's hard to see how genetic engineering could ever allow parents to design their children to have superior intelligence.

The biblical story of the expulsion of Adam and Eve from the Garden of Eden teaches us that human beings are condemned to physical and mental suffering that can never be fully alleviated by anything less than supernatural redemption. And yet Kass says that if biotechnology were to transform human nature, it would do so to satisfy the human dream of physical and mental perfection—"ageless bodies, happy souls." But how likely is that? As an indication of what he foresees, Kass says that through drugs, "we can eliminate psychic stress, we can produce states of transient euphoria, and we can engineer more permanent conditions of good cheer, optimism, and contentment." In particular, he refers to those "powerful yet seemingly safe anti-depressant and mood brighteners like Prozac, capable in some people of utterly changing their outlook on life from that of Eeyore to that of Mary Poppins." For Kass, this illustrates how biotechnology can create "happy souls."[92]

Again, I would say that Kass is exaggerating the power of biotechnology. As one looks at the scientific evidence for the genetic and neurological bases for emotional states such as happiness, fear, and depression, one must confront the natural complexity of such emotions. These emotions arise as joint products of innate temperament, life history experiences, and mechanisms of the endocrine and nervous systems interacting with one another and with contingent events in life in such complex and flexible ways that it is impossible to have any precise technical control over such emotions.

If biotechnology were ever to give parents the power to have "designer babies," parents might want to select those genetic traits that would give their children a happy temperament so that they would never become unduly sad or depressed. But how likely is that?

Fear, anxiety, and depression are associated with fluctuations in the neurotransmitter serotonin—or 5-hydroxytryptamine (5-HTT). Prozac and other antidepressant drugs are thought to work by influencing the serotonergic systems of the brain. A single gene is known to code for a protein that serves as a transporter for serotonin. This gene comes in two common forms or alleles. One allele has a short promoter region, which controls the expression of the gene. The other allele has a long promoter region. The short allele produces less of the transporter protein than does the long allele. People with the short allele are more susceptible to becoming fearful, anxious, and depressed than are people with the long allele. People tend to become depressed when they have experienced stressful events such as divorce, bereavement, or loss of a job. In one study, people who had experienced three or four such stressful events over a period of a few years were more likely to become clinically depressed if they had the short allele of the 5-HTT gene than those who had the long allele.[93]

The amygdala is one area of the brain that controls emotional experience. Using functional magnetic resonance imaging (fMRI), scientists have observed that if people are shown pictures of faces expressing fear, people with one or two copies of the short allele of the 5-HTT gene show greater neuronal activity in the right amygdala than do people with the long allele.[94] This suggests that differences in the 5-HTT gene create differences in the sensitivity of the neuronal activity in the amygdala for supporting the emotional experiences of fear, anxiety, and depression in response to stressful events. As one researcher put it, the people with the short alleles "take things too seriously."

Notice the natural complexity in all of this. Some people may be innately presupposed to "take things too seriously," and they will be susceptible to unhealthy depression. But this is only a susceptibility, and therefore it depends on whether people experience many deeply stressful life events as to whether the susceptibility to depression is expressed as debilitating depression or not. There is a complex interaction between genes, the brain, and the environment of an individual's life history. Consequently, there is no "depression gene" that parents might want to eliminate from their child's genome. There is only some genetic predisposition that will be diversely expressed depending upon life experience.

We can also see here how the capacity for emotions such as fear and anxiety is naturally adaptive. When people experience a stressful event like a divorce or the loss of a job, feeling fear or anxiety might be a necessary emotional signal to motivate them to examine their lives and look for some appropriate response to their problems. But what is naturally fitting or normal as an emotional response to stress is a matter of degree. Being either too sensitive to stress or being utterly unresponsive to stress would be unhealthy. What we need, as Aristotle observed, is an emotional balance—a golden mean—between excess and deficiency. Aristotle thought most of the virtuous traits of character required such a mean. But then it becomes impossible for parents who would genetically "design" their babies to manipulate the genetic basis for natural temperament in any precise way. Because what we need is not to excise a bad gene or insert a good gene but to achieve a balance between an appropriate genetic temperament interacting with life experiences to produce behavior that will promote a person's long-term well-being or flourishing.

If we can't rely on genetic biotechnology to give us "happy souls," can we turn to psychoactive drugs to do this, as Kass suggests? In Aldous Huxley's *Brave New World*, people who felt a little anxious or sad could take the drug soma, which induced blissful euphoria and allowed them to "escape from reality" for long periods without any painful aftereffects. Today, the selective serotonin reuptake inhibitors (SSRIs)—such as Prozac, Paxil, and Zoloft—are widely advertised on television as if they were the equivalent of soma.

Psychiatrist Peter Kramer—in his best-selling book *Listening to Prozac*—described patients using Prozac who were not just cured of depression but so transformed in their personalities as to be "better than well." Shy, quiet people were apparently turned into ebullient and social engaging people. "Like Garrison Keillor's marvelous Powdermilk biscuits," Kramer reported, "Prozac gives these patients the courage to do what needs to be done." This was the beginning, he concluded, of "cosmetic psychopharmacology," by which people could use chemicals to take on whatever personality they might prefer.[95]

But as even Kramer has conceded, this chemical transformation in personality appears to work well with only a minority of the people taking Prozac. And in recent years, many researchers—such as Peter Breggin and David Healy—have warned that the clinical evidence does not support the enthusiastic claims for SSRIs made by the drug companies selling them.[96] There have been increasing reports of many harmful side effects. This is to be expected, because all psychotropic drugs disrupt the normal functioning of the brain, and the brain responds by countering the effect of the drug, which then induces harmful distortions in the neural system. Specifically, Prozac blocks the normal removal of the neurotransmitter serotonin from the space between nerve cells. This creates an overabundance of serotonin, and the brain responds either by reducing receptivity to serotonin or by reducing the production of serotonin. Thus, the brain creates an imbalance in response to the disruption of the drug, and the brain cannot function normally. There is also growing evidence that Prozac and other SSRIs do not actually cure depression, because the antidepressant effects of these drugs are not much greater than what occurs when people are taking a placebo pill.

The fundamental problem with drugs like Prozac is one that it shares with all psychotropic drugs (including the old-fashioned ones like alcohol). Emotional suffering is a capacity of human nature shaped by evolutionary history for an adaptive purpose. Emotional suffering is almost always a signal that something is wrong in our lives. It alerts us that there is some problem either in our internal lives, in our social relationships, or in our external circumstances. A psychotropic drug does not help us to understand or solve the problem. Rather, the drug deadens the emotional response of our brain without changing the problem that provoked the emotional response in the first place. When we feel bad because of a problem in our lives, taking a psychotropic drug to make us feel better is evasive and self-defeating. As mature adults, we can understand this in the case of old drugs like alcohol. The same lesson applies to the newest drugs of the mind like Prozac.

As the Bible wisely teaches us, human life is full of suffering. To face up to that suffering, we need the moral and intellectual virtues that sustain good character. But if the Bible is correct, we can never fully satisfy our deepest longings until we are reunited to our Creator. There is no reason to believe that biotechnology will ever give us the technical means to so radically change our human nature that our souls would never feel such suffering and such longing.

Natural Ends

Considering Kass's respect for the moral wisdom of the Bible, one might expect that as chairman of the President's Council on Bioethics, he would invoke the Bible's moral teaching as a guide for the council's deliberations about the moral issues surrounding biotechnology.[97] But as anyone who has examined the work of the council can see, Kass has supervised the meetings and reports of the council so that there are almost no references to religion at all.[98] He has done this because he knows as well as anyone that many people do not accept the Bible as morally authoritative, and those who do often disagree in their interpretation of the Bible's moral teaching. Consequently, he has had to look for some common moral ground shared by citizens in a liberal democracy who belong to diverse moral and religious communities.

He has found that common ground in the natural human desires implanted in human nature by natural selection in evolutionary history. At least implicitly, he has had to appeal to the idea that the good is the desirable, and so the natural goods for human beings conform to their natural desires. He and the other members of the council can then deliberate about the wisdom of using biotechnology to try to satisfy these natural desires.

This is clear, for example, in the council's report *Beyond Therapy: Biotechnology and the Pursuit of Happiness.*[99] This report is organized as a moral assessment of biotechnology's likely success or failure in satisfying the deep human desires for healthy and happy children, for superior performance in athletic competition, for the preservation of life, and for happiness as the comprehensive desire for a flourishing life.[100] The report assumes that people will want to use biotechnology to try to satisfy such desires.

The critical question for the report is whether the likely uses of biotechnology will really satisfy these desires or not. A recurrent theme in the assessment is the "Midas problem": the possibility that when people get what they wish for, they might discover that it is not what they really wanted after all.[101] We might discover, for example, that taking psychotropic drugs so that we never feel sad would give us a shallow life that would not be worth living. In our use of biotechnology, as in our moral lives generally, we can be mistaken about our desires and desire something that turns out to be undesirable. Perhaps the biggest mistake of all would be to desire to use biotechnology in ways that would extinguish our very identity as human beings. This would be the ultimate cost: "Getting what we seek or think we seek by no longer being ourselves."[102]

The report assumes that our natural desires belong to our species as shaped by natural evolution. Our desires have been formed by natural selection over evolutionary history to promote survival and reproduction.[103] Knowing this should make us cautious about using biotechnology to radically change our evolved nature. "The human body and mind, highly complex and delicately balanced as a result of eons of gradual and exacting evolution, are almost certainly at risk from any ill-considered attempt at 'improvement.'"[104]

Thus, the report of Kass's council assumes that since our natural desires provide our ultimate motivations for action, we can assess the uses of biotechnology by how well they satisfy those natural desires. Reasoning about these desires of our human nature invokes a natural standard of judgment that is comprehensible by natural reason without any need for biblical guidance. In the moral debate over biotechnology, people on both sides of the debate must ultimately appeal to natural human desires as the ground of their arguments.

Fukuyama insists that we can regulate biotechnology by appealing to human nature so that we can promote what is naturally desirable for human beings while discouraging what is unnatural. And yet he is vague about the content of human nature. He rejects my claim that human nature is constituted by twenty natural desires. "Such lists," he claims, "are likely to be controversial; they tend either to be too short and general, or overly specific and lacking in universality."[105] What we need to know, he says, is the "Factor X" that makes human beings unique in a way that gives them moral dignity. He then goes through a list of possible traits that would qualify as "Factor X": reason, language, consciousness, moral choice, human emotions, and other factors. But he finally concludes that what is decisive is not any one of these traits but the full gamut of traits that constitute "the human whole."[106] This is confusing, because his list of human traits corresponds closely to my list of natural desires.

I say that human beings naturally desire to care for children. And Fukuyama repeatedly speaks of parental care as a natural desire.[107] I say that human beings naturally desire social ranking. And Fukuyama stresses the striving for social recognition as a natural desire.[108] I say that human beings naturally desire political rule. And Fukuyama agrees that human beings are political by nature.[109] In fact, all of the twenty natural desires on my list appear in Fukuyama's account as elements of human nature.

People on the other side of this debate—such as Silver and Stock—must also appeal to these natural desires to support their moral arguments. With respect to the moral ends of reproductive technology, Silver sometimes exaggerates in suggesting that the traditional motivations for human behavior will be completely transcended. And yet the plausibility of Silver's libertarian position depends on his implicit assumption that the new reproductive technology will be guided by the same natural desires that have always moved human beings. In particular, Silver stresses the natural desire for parental care. He repeatedly speaks of the "desire to have a child" as a "natural instinct" or "essential human desire" that has been shaped by evolutionary history as an enduring trait of human nature.[110] Although biotechnology will provide us with new means to satisfy this desire, the end is still set by our parental desire to produce and care for our children in ways that enhance their health and happiness. He argues, therefore, that the technological means should be judged good as long as they serve good ends—the satisfaction of our natural desire for children.

Silver rejects the society depicted in Huxley's *Brave New World* not only because the World State enforces eugenics through governmental coercion, but also because it has abolished marriage and parenthood.[111] Using biotechnology to manipulate the reproductive process should not bother us, he insists, as long as this has been freely chosen by parents to satisfy their natural desire for parental care of children. Of course, we can properly intervene as a society to prevent parents from inflicting obvious harm on their children. But we assume that except in rare cases of parental abuse or neglect, the natural love that most parents have for their children will move them to act for the best interests of the children. Parents can make mistakes and harm their children unintentionally. But they will learn from their mistakes and correct them as best they can. A free society with free markets will allow parents to learn by trial and error. So when Silver speaks of biotechnology as giving us "the power to change the nature of humankind," he is exaggerating, because he must assume that the natural desire for parental care will continue to direct human reproduction and child care just as it has throughout human history.[112]

Like Silver, Stock suggests at first that "redesigning humans" through genetic manipulation will include a redesigning of their fundamental motivations. But then he pulls back from this exaggerated assertion, and argues instead that choices about using reproductive technology to change the human germline will be made by parents moved by the same natural desires that have always been part of human nature. He writes: "To figure out which traits we will want for our children once we have the power to make such choices, we must think long and hard about who we are. Our evolutionary past speaks to us through our biology and fashions our underlying desires and drives." He then lists some of those "desires and drives" instilled in us by Darwinian evolution. His list includes sexual mating, parental care, familial bonding, status, power, wealth, and beauty.[113] He thus appeals to those same natural desires rooted in our Darwinian human nature that I have identified.

The persuasiveness of Stock's libertarian argument depends on his implicit claim that although parents will make mistakes if they are free to choose how to employ new reproductive technologies, we can generally rely on their commonsense judgment, because they will be guided by those natural desires that have always constituted the ground for moral experience. If modification of the human germline arises from parental choices about those improvements that parents want for their children, we can trust, Stock says, that such enhancement will fall "within the range of normal human performance."[114] If so, then it would seem that Stock's project for "redesigning humans" has not abolished human nature after all. As long as the human beings using biotechnology do so in the service of their natural desires, their technical means might be new, but their moral ends will be rooted in the enduring desires of human nature.

Conclusion

The moral debate over biotechnology shows us that moral judgment depends ultimately on natural human desires rooted in human biological nature. The good is the desirable, and thus moral controversy requires deliberation about how best to satisfy our desires harmoniously over a whole life well lived. Some people worry, however, that this moral reliance on natural desires cannot work once we have the power to use biotechnology to abolish our human nature. To prevent the "abolition of man," they believe, we must invoke the moral teaching of biblical religion to teach us that we must limit our biotechnological mastery of nature to respect the limits set by God as our Creator.

But although the Bible can reinforce our natural moral sense, the Bible cannot stand alone as a moral guide because it often lacks moral authority, moral clarity, and moral reliability. It lacks moral authority with those people who doubt that it is truly a revelation from God. It lacks moral clarity because its moral teaching is often too vague to give us precise moral instruction. And it lacks moral reliability because some of its teachings are immoral and therefore need to be corrected by our natural moral sense.

Even in an age of biotechnology, we can rely on our natural desires as a moral guide that stands independently of the Bible or any other religious text. The power of biotechnology over human nature has been exaggerated both by its unduly fearful opponents and by its unduly hopeful proponents. Biotechnology will be limited both in its technical means and in its moral ends. It will be limited in its technical means because it will be constrained by the adaptive complexity of human nature. And it will be limited in its moral ends because it will be used to satisfy the natural desires of human nature.

If we could give up both the utopian hopes and the apocalyptic fears evoked by biotechnology, then we could begin a more sober moral deliberation about how we might put our new technological powers into the service of our natural desires.

Notes

1. See, for example, John Hare, *Why Bother Being Good? The Role of God in Ethics* (Downers Grove, Ill.: Intervarsity Press, 2002).

2. Larry Arnhart, *Darwinian Natural Right: The Biological Ethics of Human Nature* (Albany: State University of New York Press, 1998).

3. Various people have made this argument. See, for example, Carson Holloway, *Darwinism and Political Theory* (Dallas: Spence Publishing, 2004).

4. Haldane's lecture has been reprinted in Krishna Dronamraju, ed., *Haldane's Daedalus Revisited* (Oxford: Oxford University Press, 1995).

5. Ibid., 37, 48–49.

6. Aldous Huxley, *Brave New World and Brave New World Revisited* (New York: Harper and Row, 1965).

7. C. S. Lewis, *The Abolition of Man* (New York: Macmillan, 1947), 86.

8. George W. Bush, Address to the Nation, August 9, 2001, in *The Future Is Now: America Confronts the New Genetics*, ed. William Kristol and Eric Cohen (Lanham, Md.: Rowman and Littlefield Publishers, 2002), 306–310.

9. Ibid., 309.

10. Leon Kass, *The Beginning of Wisdom: Reading Genesis* (New York: Free Press, 2003).

11. Ibid., 1.

12. Ibid., 3.

13. Ibid., xiv.

14. Ibid., 16.

15. Ibid., 44.

16. Francis Bacon, *Francis Bacon: The Major Works*, ed. Brian Vickers (Oxford: Oxford University Press, 2002), 125–126.

17. Charles Darwin, *The Origin of Species*, 6th ed., in *The Origin of Species and The Descent of Man* (New York: Random House, Modern Library, 1936), 2.

18. Ibid., 373.

19. See David Novak, *Natural Law in Judaism* (Cambridge: Cambridge University Press, 1998).

20. Romans 1:20, 2:14–15.

21. See Robert A. Greene, "Instinct of Nature: Natural Law, *Synderesis*, and the Moral Sense," *Journal of the History of Ideas* 58 (April 1997): 173–194.

22. See Larry Arnhart, "Thomistic Natural Law as Darwinian Natural Right," in *Natural Law and Modern Moral Philosophy*, ed. Ellen Frankel Paul, Fred Miller, and Jeffrey Paul (Cambridge: Cambridge University Press, 2001), 1–33.

23. Kass, *Wisdom*, 173–174, 193–196.

24. Ibid., 98–105, 111–120.

25. Ibid., 112–113.

26. See ibid., 427–432.

27. James Madison, Alexander Hamilton, and John Jay, *The Federalist* (New York: Random House, Modern Library, 1937), 230.

28. Jeremy Rifkin, *Who Should Play God? The Artificial Creation of Life and What It Means for the Future of the Human Race* (New York: Delacorte Press, 1977).

29. Jeremy Rifkin, *Algeny* (New York: Viking, 1983), 252.

30. Bacon, *Major Works*, 124–26, 147–148, 191–192.

31. See John Houck and Oliver Williams, eds., *Co-Creation and Capitalism: John Paul II's* Laborem Exercens (Washington, D.C.: University Press of America, 1983); Philip Hefner, *The Human Factor: Evolution, Culture, and Religion* (Minneapolis: Fortress Press, 1993); Hefner, *Technology and Human Becoming* (Minneapolis: Fortress Press, 2003); and Ted Peters, *Playing God? Genetic Determinism and Human Freedom*, 2nd ed. (New York: Routledge, 2003).

32. Robert Alter, *The Five Books of Moses: A Translation with Commentary* (New York: Norton, 2004), 58–59. Except when I am quoting from Alter's translations of the first five books of the Bible, all of my other biblical quotations are from *The New Jerusalem Bible* (Garden City, N.Y.: Doubleday and Company, 1985).

33. Kass, *Wisdom*, 221, 224–225, 231–232.

34. Ibid., 242–243.

35. Ibid., 236, 239.

36. See, for example, ibid., 10, 60, 90, 114, 118–119, 130–131, 138–139, 146, 157, 173, 232, 242, 249, 293–294.

37. Alter, *Five Books of Moses*, 58.

38. Kass, *Wisdom*, 130.

39. On the importance of agriculture and the domestication of plants and animals for the emergence of civilization, see Jared Diamond, *Guns, Germs, and Steel: The Fates of Human Societies* (New York: Norton, 1997); and A. M. T. Moore, G. C. Hillman, and A. J. Legge, *Village on the Euphrates: From Foraging to Farming at Abu Hureyra* (Oxford: Oxford University Press, 2000).

40. See Brian Alexander, *Rapture: How Biotech Became the New Religion* (New York: Basic Books, 2003), 22–24.

41. Kass, *Wisdom*, 249.

42. See, for example, Bacon, *Major Works*, 151–152, 266–270, 471.

43. See Alfred North Whitehead, *Science and the Modern World* (New York: Macmillan, 1926), 18–19; Stanley Jaki, *The Savior of Science* (Edinburgh: Scottish Academic, 1990); and David C. Lindberg, *The Beginnings of Western Science* (Chicago: University of Chicago Press, 1992).

44. Pope John Paul II, *The Gospel of Life:* Evangelium Vitae (New York: Times Book, 1995), 77–78, 108.

45. See Elliot N. Dorff, *Matters of Life and Death: A Jewish Approach to Modern Medical Ethics* (Philadelphia: The Jewish Publication Society, 1998); and Dorff, "Stem

Cell Research—A Jewish Perspective," in *The Human Embryonic Stem Cell Debate: Science, Ethics, and Public Policy*, ed. Suzanne Holland, Karen Lobacqz, and Laurie Zoloth (Cambridge: MIT Press, 2001), 89–93.

46. Leon Kass, *Life, Liberty and the Defense of Dignity* (San Francisco: Encounter Books, 2002), 257–259.

47. Peters, *Playing God?* 191–192.

48. Dorff, *Life and Death*, 396–400.

49. For a good survey of the various attempts by Jewish and Christian commentators to morally justify the binding of Isaac, see Ronald M. Green, *Religion and Moral Reason* (New York: Oxford University Press, 1988), 77–129.

50. Kass, *Wisdom*, 344–345.

51. Ibid., 333, 351.

52. Søren Kierkegaard, *Fear and Trembling and The Sickness Unto Death*, trans. Walter Lowrie (Princeton: Princeton University Press, 1954), 64, 69–70, 86–87.

53. Ibid., 82.

54. Ibid., 49.

55. Immanuel Kant, *The Conflict of the Faculties*, trans. Mary J. Gregor (New York: Abaris Books, 1979), 115.

56. Green, *Religion and Moral Reason*, 82.

57. Thomas Aquinas, *Summa Theologica*, I–II, q. 100, a. 8, ad. 3.

58. Green, *Religion and Moral Reason*, 101.

59. Alter, *Five Books of Moses*, 978.

60. Ibid., 843.

61. See Kass, *Wisdom*, 492–494.

62. Robert Alter, *The David Story: A Translation with Commentary of 1 and 2 Samuel* (New York: Norton, 1999), xviii.

63. Niccolo Machiavelli, *The Prince*, chapters 6, 13, 26.

64. See E. N. Elliott, ed., *Cotton Is King, and Pro–Slavery Arguments* (Augusta, Ga.: Pritchard, Abbott and Loomis, 1860), viii–xi, 337–380, 459–521, 811–816, 841–877; and Edward Westermarck, *Christianity and Morals* (London: Kegan Paul, Trench, Trubner and Company, 1939), 282–306.

65. See John Locke, *Second Treatise of Government*, secs. 6, 22–24; and Jonathan Conrad, "Locke's Use of the Bible," PhD dissertation, Northern Illinois University, 2004.

66. Lewis, *Abolition of Man*, 56, 97.

67. Thomas Aquinas, *Summa Theologica*, I–II, q. 100, a. 1.

68. See Walter Burkert, *Creation of the Sacred: Tracks of Biology in Early Religions* (Cambridge: Harvard University Press, 1996); Vernon Reynolds and Ralph Tanner,

The Social Ecology of Religion (New York: Oxford University Press, 1995); and David Sloan Wilson, *Darwin's Cathedral: Evolution, Religion, and the Nature of Society* (Chicago: University of Chicago Press, 2002).

69. Here and elsewhere in this chapter I have incorporated some passages from my article "Human Nature Is Here to Stay," *The New Atlantis* 2 (summer 2003): 65–78.

70. Lee Silver, *Remaking Eden: How Genetic Engineering and Cloning Will Transform the American Family* (New York: Avon, 1998).

71. Ibid., 15, 71, 105, 153, 231, 281.

72. Ibid., 273–277.

73. Ibid., 266, 274.

74. Ibid., 9–13, 144, 187, 253–255, 295, 307–308.

75. Ibid., 11.

76. Gregory Stock, *Redesigning Humans: Our Inevitable Genetic Future* (Boston: Houghton Mifflin Company, 2002), 3–4, 64, 76–77, 111.

77. Ibid., 64–70, 105–109, 135.

78. Bill McKibben, *Enough: Staying Human in an Engineered Age* (New York: Times Books, 2003), 44–65.

79. Ibid., 208–209.

80. Ibid., 29, 99.

81. Leon Kass, "Preventing a Brave New World," in *The Future Is Now: America Confronts the New Genetics*, ed. William Kristol and Eric Cohen (Lanham, Md.: Rowman and Littlefield Publishers, 2002), 225.

82. Francis Fukuyama, *Our Posthuman Future: Consequences of the Biotechnology Revolution* (New York: Farrar, Straus, and Giroux, 2002), 82–83.

83. Francis Bacon, *The New Organon*, ed. Fulton H. Anderson (Indianapolis: Library of Liberal Arts, 1960), I.3–4, 39.

84. Leon Kass, *Toward a More Natural Science* (New York: Free Press, 1985), 151–153.

85. For the history of biotechnology, see Robert Bud, *The Uses of Life: A History of Biotechnology* (Cambridge: Cambridge University Press, 1993); and James D. Watson, *DNA: The Secret of Life* (New York: Knopf, 2003). I have surveyed some of this history in my article "Biotech Ethics" in *The Encyclopedia of Science, Technology, and Ethics*, ed. Carl Mitcham, 4 vols. (New York: Macmillan Reference, 2005).

86. See McKibben, *Enough*, 25–26; and M. J. Chorney et al., "A Quantitative Trait Locus Associated with Cognitive Ability in Children," *Psychological Science* 9 (May 1998): 159–166.

87. Linzy Hill et al., "A Quantitative Trait Locus Not Associated with Cognitive Ability in Children: A Failure to Replicate," *Psychological Science* 13 (November 2002): 561–562.

88. Robert Plomin, "Genetics, Genes, Genomics and g," *Molecular Psychiatry* 8 (2003): 1–5.

89. See Howard Gardner, *Frames of Mind: The Theory of Multiple Intelligences* (New York: Basic Books, 1983).

90. On "emotional intelligence," see Antonio Damasio, *Descartes' Error: Emotion, Reason, and the Human Brain* (New York: G. P. Putnam's Sons, 1994); Gareth Matthews, Moshe Zeidner, and Richard D. Roberts, *Emotional Intelligence: Science and Myth* (Cambridge: MIT Press, 2002); and Marlene Sokolon, *Political Emotions: Aristotle and the Symphony of Reason and Emotion* (DeKalb: Northern Illinois University Press, 2005). Even in the field of economics, which has long been dominated by "rational choice theory," there is now growing research on the emotional basis of decision-making. See Paul W. Glincher, *Decisions, Uncertainty, and the Brain: The Science of Neuroeconomics* (Cambridge: MIT Press, 2003); Colin F. Camerer, "Strategizing in the Brain," *Science* 300 (June 13, 2003): 1673–1675; and Alan G. Sanfey et al., "The Neural Basis of Economic Decision-Making in the Ultimatum Game," *Science* 300 (June 13, 2003): 1755–1758.

91. Kass, *Wisdom*, 38.

92. Leon Kass, "Ageless Bodies, Happy Souls: Biotechnology and the Pursuit of Perfection," *The New Atlantis* 1 (Spring 2003): 12.

93. See Constance Holden, "Getting the Short End of the Allele," *Science* 301 (July 18, 2003): 291–293; and Avshalom Caspi et al., "Influence of Life Stress on Depression: Moderation by a Polymorphism in the 5-HTT Gene," *Science* 301 (19 July 2003): 386–389.

94. See Ahmad R. Hariri et al., "Serotonin Transporter Genetic Variation and the Response of the Human Amygdala," *Science* 297 (July 19, 2002): 400–403.

95. Peter Kramer, *Listening to Prozac* (New York: Viking, 1993), x, 10–11, 15.

96. Peter Breggin, *The Anti-Depressant Fact Book* (Cambridge, Mass.: Perseus Publishing, 2001); and David Healy, *Let Them Eat Prozac: The Unhealthy Relationship Between the Pharmaceutical Industry and Depression* (New York: New York University Press, 2004).

97. I have written a general account of the council's work in "The President's Council on Bioethics," in *The Encyclopedia of Science, Technology, and Ethics*, ed. Carl Mitcham, 4 vols. (New York: Macmillan Reference, 2005).

98. Reports of the President's Council on Bioethics and transcripts of the council's meetings can be found at http://www.bioethics.gov.

99. Leon Kass, ed., *Beyond Therapy: Biotechnology and the Pursuit of Happiness*, A Report of the President's Council on Bioethics (New York: Dana Press, 2003).

100. Ibid., xxv, xxxi, 24–25.

101. Ibid., xxv, 177, 207, 264, 314.

102. Ibid., 177.

103. Ibid., 102, 227, 247, 254–255, 278, 323.

104. Ibid., 323.

105. Fukuyama, *Our Posthuman Future*, 139.

106. Ibid., 150, 171, 174, 176.

107. Ibid., 99, 141–142, 187.

108. Ibid., 44–46, 64–67, 117, 149.

109. Ibid., 44–46, 149, 164–166, 170, 186.

110. Silver, *Remaking Eden*, 81–83, 169, 229, 295, 308.

111. Ibid., 9–10.

112. Ibid., 13, 273.

113. Stock, *Redesigning Humans*, 116–123.

114. Ibid., 116.

SEVEN

A Transcendent Vision

Theology and Human Transformation

RICHARD SHERLOCK

I am honored to be included in this collection of essays on biotechnology and I hope to fulfill a role as agent provocateur in support of the indispensable role of theology in the discussion of biotechnology. This will be especially true in the case of human biology as distinct from agricultural, food, and animal applications. I do not represent all versions of Christian theology.[1] The differences among theological traditions are as great as those between philosophical traditions in the West. I shall articulate a view of theology, and what it may teach, that I believe is broadly acceptable to a number of Christian theological traditions, though I claim neither comprehensiveness nor prophetic authoritativeness.

Christian theology is, in my view, an enterprise of the believing community. Theology seeks to articulate a coherent set of communal beliefs about God, humanity, human destiny, and our moral lives as part of the people of God. Thus, the fundamental test of theology is coherence with those core convictions that the community holds as precious or sacred.[2] These convictions are, I believe, those that some philosophers call "control beliefs," or beliefs held by the community that makes the community distinctively what it is.[3] They control what members of the community find as rational answers to the most important human questions. Such beliefs may be analogized to what Aristotelian philosophers might call the essential properties of the object. The essential properties of an object P', P", P"' . . . are such that if any of these properties are missing the object is no longer what it was.[4] This analogy may be helpful but we should constantly remind ourselves that theology is an intra-communal activity, not an activity of neutral or detached investigation. In this respect, however, theology is no different than any other investigation. All of us, theists or nontheists, biblical literalists or narrative theologians, have background or control beliefs that shape all of our other beliefs about the world around us.

Let me provide a recent example. In her wonderful book *The Sacred Depths of Nature*, the distinguished biologist Ursula Goodenough rearticulates the "religious naturalism" of Emerson and Dewey for the twenty-first century.[5] Goodenough may be described as a friendly nontheist. At the outset and throughout the book she announces her lack of any theistic belief. She simply cannot tell us to believe in any divine being.[6] Later on, however, she discusses "religious" or "mystical" experiences. As befitting the personal tone of the book, she readily admits to having had such experiences and to their transformative, awesome power. For her, however, they simply cannot be experiences of a Divine Other that touches the human soul. Nontheism is her control belief. Hence, these sorts of experiences must be understood as simply "wondrous mental phenomena."[7]

Theology, in my view, is not an enterprise that tries to make Christian beliefs acceptable to what used to be called "modern man." Rather, theology's task is to articulate faith, not to make it into something acceptable to Enlightenment rationalism. There are technical reasons for theology's preference for coherence epistemology. Theology bears witness to what the community knows as true. The believer most often says, "I know P to be true," where P is a theological belief such as "Jesus Christ is the savior of humanity." Coherence epistemology, where the truth of a belief depends on the coherence of our other beliefs, is the only way to a robust account of truth such as found in a statement like "I am certain that Jesus Christ is my savior."[8] The alternative to coherence epistemology, correspondence epistemology, will not get us to certainty, only probability. Correspondence epistemology holds that the truth of a proposition is its correspondence with an external or mind independent state of affairs. But since the human mind never has direct access to the external world, the most that one can claim is that P probably corresponds to an external state of affairs because P is consistent with what others hold and because P is persistent to us through time.

The Christian faith that theology articulates is grounded in what Christians understand as revelation in word and deed. Christians revere and hold sacred the scriptural witness but not uniformly. In my view Christians must read all of scripture in light of the Jesus narrative in the New Testament.[9] Most of the rest of the New Testament constitutes theological reflections on the Jesus narrative. The moral teachings of the great prophetic voices of the Old Testament / Hebrew Bible are held as normative because they cohere with the moral voice of the gospel narrative. If there are principles illuminated in the historical books of the Hebrew Bible that Christians should revere it is because, and only because, they can be implicitly found in the New Testament. Likewise, Christians will accept some teachings from the Torah such as the Decalogue because the principles that are stated there are confirmed by the Jesus narrative. For example, Christians should accept the *Akedah*, the story of the binding of Isaac, because it teaches in vivid form a deeply Christian message. First, it teaches fear of God that is the ground of God's mercy, and it especially teaches absolute trust in God.[10] This latter trust in God is, I submit, the core of the teaching of Jesus and is seen most especially in his death and resurrection. Whether we "consider the lilies of the field" or the "birds of the air" Christianity begins with a trust in God over and against all the powers of the cosmos. The great example of this trust is Jesus whom Christians know as savior or by the titles of "son of God," "coeternal Word," "Christ," and even "son of man."[11] The New Testament portrays an entirely human Jesus in the hours before his death. He prays three times to a divine father asking, "If it could be possible is there another way than my crucifixion? If not, I place my trust and my life in your hand, Father."[12] It is, I think, easy to see why Christians saw in the crucifixion a sort of replay of Abraham and Isaac. The same absolute trust is seen and the same witness to God's mercy and love is exemplified. Christian faith contains a moral teaching about absolute love of neighbor, a trust in and love of God, and a deep conviction of His perfect love of us. These are all core Christian teachings. But the absolute foundation of Christian faith is the certainty that the ancient Apostles' Creed is true: He rose on the third day. The precise nature of this rising has been the subject of much dispute but for my purposes it can be narrowed down to the basics. Whatever properties made Jesus of Nazareth the distinct individual he was from birth to death are now the properties that continue to set apart a distinct being Jesus Christ, which continues to exist after the crucifixion of Jesus. This resurrected Jesus may have additional qualities added on through time just as I now have many qualities that I once did not have, for example, my figure is more rotund, my beard is gray, and I require glasses. But whatever one wants to argue is the source of my personal identity remains from the time I first existed till now. So too with Jesus of Nazareth, who is the same person now as risen Lord.[13]

The moral teaching of the Christian faith puts love of neighbor first among all the principles of virtuous action that one might conceive of. The love command requires complete commitment to the welfare of others, especially those others who are imperfect, weak, or handicapped in various ways, such as the prisoners, the poor, the weak, and the hungry of Matthew or the outcast and despised of the story of the good Samaritan.[14] Pure love forbids anger, violence, or hatred of others. Love requires active nurturing of the other, especially, again, the weak and the despised.[15]

Love of neighbor is possible because of our commitment to God and His love of us. A covenant with and complete trust in God is the basis of the

Christian sense of humility before the Divine. We are not our own makers nor are we our own saviors. Commitment to neighbor and God, I believe, means that Christian faith must find itself in opposition to the rational egoism of modern rational choice theory in all its forms, including social contract theory, which is its real ancestor. Likewise Christians must reject the classical attraction to so called great-souled individuals whose greatness is displayed in founding cities or in triumphs on the battlefield. Aristotle's fascination with such individuals in the *Nicomachean Ethics* reflects a fascination with human versions of greatness that Christians must reject decisively in favor of God's glory.[16] The kingdoms of Solomon, Alexander, and Augustus are not to be revered, only His.

My statement here has been only one among several theological positions found in the Christian tradition. I am prepared to defend it in depth, but this is not the proper forum. What I have argued is that Christian faith teaches awe before, love of, and trust in God as the beginning of virtue and wisdom. This triad is grounded in God's love of us and His providing for our eternal destiny. His care for our welfare frees us from such care so that we can care for others above all else.

Theology and Biotechnology

Concerning the relation between theology and biotechnology three fundamental questions must be addressed. First, is a theological version of what Larry Arnhart calls a "cosmic teleology" required to properly comprehend biotechnology? Second, what might a Christian theological witness such as mine have to say about developments in biotechnology? Thirdly, what should be the theological voice in the public discussion of biotechnology?

Whether one believes that a transcendent vision is required to fully comprehend the future of biotechnology depends on whether one accepts as even possible the claims of fervent advocates to be able to drastically alter human nature itself. If one regards this as even possible, the moral tradition of the West may no longer be a solid ground from which to comprehend the proper goals of and limits to biotechnology.

The Western moral tradition, since the ancients, has been heavily invested in grounding morality in a view of human nature as stable and morally charged. There are a wide variety of claims about human nature in this tradition from Aristotle's teleological biology of human flourishing through the Thomistic and neo-Thomistic attempts to appropriate Aristotle in behalf of a Christian natural law, to the insistence of Hume, Smith, and Darwin of a natural moral sense or sentiment to newer attempts to ground morality in human nature biologically conceived such as those of Arnhart and William Casebeer.[17]

Does the teleological form of the argument, whether in Aristotle, St. Thomas, or others present a similar structure? Human beings are said to be creatures of a certain sort, such that they act in certain ways to attain happiness, which is equivalent to saying that they flourish as the kind of beings they are by acting in certain ways. Moral principles arise out of this sort of biological morality. For example, it can be argued that human beings must survive in groups as social animals, as Aristotle's famous "political animals." As such we must develop rules of group cooperation such as honesty or trustworthiness, which are necessary if we are to maintain the group, which sustains our existence. A second example follows from our sexual nature. Human beings biologically reproduce by sexual union. From this arises the requirement of parental investment of time in the nurture of children. This investment is unequally distributed not only because of social conditioning or role stereotyping but also because in evolutionary time the female began as the primary food source for the infant and males as both the primary source of physical security for the group and as hunters, the primary source of food for the mother. Hence we develop over time moral beliefs or principles that articulate ways of life that we regard as right because they give support to what nature initially teaches: parents must care for their children, spouses have responsibilities to care for each other, and divorce is generally wrong.

Moral sense theorists have a different and less teleological view of our moral nature. Hume and his followers, including Darwin, do not argue that there are biological propensities or "natural desires" the fulfillment of which leads to happiness or human flourishing. Rather these theorists hold that we have a natural sense of moral approval and disapproval, which is expressed as a sense of pleasure. The relationship is not sequential. We do not feel pleasure or happiness in contemplating a particular act because antecedently we believe it to be right, rather, it is in feeling the pleasure we feel that it is right.[18]

From the naturalist's point of view, it would seem that biotechnology and allied chemical technologies that alter feelings call into serious question the reliance on human nature as a ground of moral claims. Just at the time that "naturalized' moral theory is making a serious comeback among moral theorists, science seems ready to undermine it. At the very least this appears to be the result of the claims of the promoters of what I shall call "strong biotechnology." This is the thesis that we are on the verge of vast transformations of human nature, a thesis that is shared by friend and foe alike. Promoters write expansive treatises about "redesigning human nature" or creating immortal beings while critics respond with either careful warning about so doing or alarmist screeds. Alarm, however, is not a proper response to what one regards as impossible. Rational people are not alarmed by the prospect of an invasion from Mars. Nor should we waste time arguing against doing what we simply can't do. The critics seem, therefore, to accept the premise of the promoters: vast transformations of human nature are possible.

The most frequently cited promoters are biologist Lee Silver of Princeton and bioethicist Greg Stock of UCLA. Silver's *Remaking Eden*, written in the aftermath of Dolly, was among the first explicit statements of "strong biotechnology."[19] Stock's somewhat later *Redesigning Humans* provides a very useful compendium of the claims of the redesign thesis.[20] These are, however, hardly the only writers who promote such claims.[21] For more material one might consult the organization betterhumans.com, which provides information and columns for newspapers and journalists on the theme of their name. One might also read the work of a number of those associated with ideas to vastly extend human longevity such as Stanley Shostak, who writes of immortality via biotechnology[22] or geneticist Aubrey de Grey of the University of Cambridge, who is the founder of the Methuselah mouse prize.[23] The umbrella organization for many of these thinkers is the World Transhumanist Association chaired by philosopher Nicholas Bostrom of the University of Cambridge.[24] Its Transhumanist Declaration provides a useful short statement of the strong biotechnology position. This is not the place for a full analysis of transhumanism, but the first of the seven points of the declaration bears quoting here:

> Humanity will be radically changed by biotechnology in the future. We foresee the feasibility of redesigning the human condition including such parameters as the inevitability of aging, limitations on human and artificial intellects, unchosen psychology, suffering, and our confinement to planet earth.[25]

Sending colonies to Mars would not involve strong biotechnology but conquering aging and even death or altering our basic psychological makeup surely would.

The problem for the Western moral tradition is that if the claims of "redesign" are even possible then it would appear that we can no longer rely on nature itself as a complete and sufficient basis of moral theory. If nature can be redesigned in ways implied in the Transhumanist Declaration, the moral principles that will be needed to guide redesign cannot be those found in the original design itself. They will need to be transnatural or transcendent principles that tell us what the cosmic purpose and context of human life is, such that we could judge whether or not the redesign might or might not further our reaching our transcendent destiny. Though not stating a theological solution such as I am doing, philosopher Langdon Winner states the problem nicely: "One serious consequence of the move to abandon a vital concern for humans and to search for more exotic, posthuman ways of being is to remove the foundations on which some crucial moral and political agreements can be sought—an appeal to our common humanity."[26]

If a baseball player uses steroids to increase his muscle mass so he can hit more home runs he may very well be "cheating," but on the specific question of whether he will be a better player we would certainly have to say yes. His ability to play the game will be improved. But that is because we know the rules of the game. If we know the rules of any sport we can know what an improved or an enhanced player is. So it is with any human activity. If we know the goal or *telos* of the activity we can know what enhancement of the person engaged in the activity is. Because we know the purpose of something like computer programming or surgery we can know what a better or improved programmer or surgeon is. However, do we have any generally accepted idea of what a better or improved human being is? The honest answer must be no. Or at least we have a whole variety of notions of improvement, none of which are even close to being generally accepted. It may very well be the case that biotechnology will radically change the human condition in the future, making the question of improvement or betterment a different question.

If, however, redesign of the sort promoted by the World Transhumanist Association is possible, then we cannot rely on nature to provide moral guidance about the nature of this transformation. For example, one biological feature of human beings is sexual reproduction by erotic coupling as distinct from asexual reproduction of some plants and microorganisms. This fact has provided the ground for much moral thinking about the nature of erotic union with a reproductive *telos*.[27] The moral standing of birth control, which undermines this *telos*, and same-sex intimacy, which denies this *telos*, are widely discussed. Yet in the age of cloning, which is specifically asexual reproduction, are not all of these moral positions that are based on the supposedly "natural" form of reproduction called into question?

Perhaps we can make this point with direct reference to the essay in this volume by Ronald M. Green. Green is a supporter of human biotechnology and even in a moderate way of germline enhancement. He wisely recognizes that scientifically we are a long way from enhancement technologies that are safe and reliable. He also properly addresses six objections to "enhancement" that are unrelated to either specific therapy or "prevention," such as genetic engineering to avoid Huntington's disease later in life. While the concerns he addresses such as safety, justice, and consent are important, Green entirely avoids the core problem: Do we know what constitutes enhancement? Just because we can say that increasing the average life span from 40 years to more than 80 years as we have done in the last century is generally a good thing it does not mean that an increase to 150 years would be similarly good. Green avoids giving us any account of the standard by which we should judge enhancement. Such a standard, like the rules of a game, will require us to know what the purpose of life is. Green might say that this is a personal choice. If so then why should public agencies fund research devoted to any enhancement program, which will inevitably be limited to a specific end such as longevity?

Green's essay is all too familiar on these issues. It is a plea to start dreaming about the possible, focusing on a series of rather tame and obvious concerns while avoiding the deepest questions that would call into question the very enterprise of enhancement. As millions go hungry and malaria and AIDS scourge continents, is the possible about which Green dreams ever the truly desirable use of resources?

If nature can no longer provide the sort of guidance that we need for these transformations, must we not appeal to a transnature or transcendent nature to evaluate such transformations? One cannot appeal to nature to determine whether nature should be altered. Nature will always counsel no. But this cannot be the right answer. War, starvation, and disease have always been part of nature. Surely if we could eliminate those features of our existence we should. In another example, many naturalists appeal to something like a human natural moral sense. Yet in the age of psychopharmacology can we not have people feel pleasure in situations where discomfort was once the most appropriate response? If as some assert, the moral sense of pleasure and discomfort is an evolutionary feature that alerts us to something that might be wrong with our life, we can now alter that sense so that what might have once been thought of as right or proper is no longer so. Again would we not now need a transcendent standard to judge whether actions that might not now arouse discomfort are wrong or right? If nature can now be altered, nature can no longer be seen as the standard from which a sound moral teaching can develop. Grant this, and the alternatives would seem to be reduced to some version of relativism, which gives up on serious moral thinking or a transcendent or theological basis of ethics that grounds morality in that which cannot change: God.

At this point we might make a provisional distinction between improvement or enhancement that I shall call "weak biotechnology" on the one hand and redesign or strong biotechnology on the other. Enhancement of some human capacities is already with us. Steroids can enhance muscle build. Growth hormone can enhance height even of persons whose height is in the normal range. Stimulants originally developed for the treatment of attention deficit disorder (ADD) and its cousin attention deficit hyperactivity disorder (ADHD) can now be used to sharpen adult mental function even though there is scant evidence that ADD or ADHD affects adults. Originally, drug treatment for depression had side effects such that they could be used reasonably only for those suffering from real depression, which does have a biological basis. But newer forms of antidepressants such as selective serotonin reuptake inhibitors (SSRIs) have few if any of these effects and can be used by otherwise normal people to feel better or happier about life in general.

In these ways improvement or enhancement is certainly possible. In the future with better brain mapping via molecular machines and with ever-smaller memory chips we certainly will be able to do much more to alter brain functions. This is not science fiction, it is work that is on the drawing board. Moore's law taken out to 2020 shows that one bit of information will be stored on one atom, and with developments in quantum computing we will get vastly improved modeling of brain function.

No serious defender of naturalism can deny what I have called weak biotechnology. What then would be redesign along the lines suggested by Silver, Stock, and the Transhumanists? They might mean something like serious but not necessarily complete alteration of the structure of human desires such as those for pleasure, offspring, or achievement. Or they might mean dramatically altering the human brain. The definition of "nature" here cannot be so vacuous that we can never know what counts as redesign. The statement that all men desire the good is at best a trivial truism until the good is specified. Then we can see whether it is the case that most persons desire the good as so specified. Arnhart's fine work tries to be specific about the general point and careful in marshaling evidence to support his conclusions. He writes of a "species specific behavioral repertoire of homo sapiens" that includes society, language, social learning, family, nurture and is moved by a proper self-love and concern for those bound to them like family.[28]

At another place he describes twenty "natural desires" that are present throughout human societies. Not that all people have them but that "our commonsense experience of the world would suggest that these twenty desires are indeed universal features of the human condition." Some of these twenty desires fall victim of the "vagueness objection," like beauty and aesthetic pleasure. Others are old chestnuts like sexual mating, child nurturance, or familial bonding.[29]

Elsewhere Arnhart is enamored of the Hume/Smith/Darwin natural moral sense theory, which, he could claim, is the basis of these twenty natural desires, the fulfillment of which will make us happy. There are a number of technical criticisms that might be made of any such list but they are not really germane to my point here. Any lists of rules or desires face the problem of a conflict of rules or desires. What is the master rule appeal to which will resolve the conflict?

I do not want this essay to be fundamentally based on wild speculation of science fiction scenarios. But I would point out that the vastly increased ability to shape the human brain in the near future and the enormous improvements made available by quantum computing are very likely to lead to the possibility of human machine hybrids such as cyborgs that would involve drastic alterations of what would otherwise be thought of as human nature.[30]

Secondly, I do not think it is wise for opponents to adopt what is a "God of the gaps" strategy. The old–fashioned God of the gap strategy worked like this: There was some incomplete feature of our understanding of nature, for example, speciation. This was the "gap" that God supposedly filled in. Since we didn't know how the various species could have developed, it is evidence that a superior being, God, must have created them. The approach by critics of the "redesign" thesis has the same structure of argument. Since we can't now do some redesign and since we cannot even conceive of how it might be done we should regard claims of redesign as more or less bogus. I think we are properly wary of such an argument. The history of science is replete with examples of things that were thought impossible until they were actually done. Redesign of any sort may be difficult and it certainly may be morally questionable, but I do not think critics should hang their hat on the notion that it can't be done.

The distinction between enhancement, which is already being done, and redesign is a fuzzy one at best.[31] We can certainly give examples at either end of the spectrum but setting forth a clear demarcation in the middle is probably impossible. In this fashion it resembles the well-known problem of distinguishing therapy and enhancement in medical technology. We can be moderately certain that germline therapy for a family's cystic fibrosis is therapy. Germline alteration of normal height to extraordinary height (say from 5'7" to 6'5") would be enhancement. But what about germline alteration to fix colorblindness or male pattern baldness? For the purposes of this essay, however, I am not convinced that the distinction matters. The terms enhancement or improvement that critics must admit are currently applicable to available technology like steroids or Prozac raise exactly the same issues as do the more expansive issues of redesign. Enhancement surely implies that we can do something better than we otherwise would. Is this not to achieve some good faster, more often, or easier than we would otherwise be able to do? I know what improved athletic performance is because I know the rules of the game. A tennis player is better if he serves harder, a basketball player would be better if he can hit a greater percentage of three-point shots. But what are the rules of life that allow me to judge that Prozac for otherwise normal people is an improvement or using germline engineering for colorblindness is an enhancement? While I regard the distinction between enhancement and redesign or improvement as not well addressed either by promoters like Silver or Bostrom or critics like Fukuyama or Arnhart, the distinction may not be crucial to the point I am making. Since enhancement and redesign are rungs on a ladder I want to know what the arguments are for, first, getting on the ladder in the first place, or, second, how many steps to take.

The critics must admit that some biotech alterations are possible through human growth hormone, psychotropic drugs, and asexual reproduction, such as cloning. Since some transformation is possible, critics can no longer appeal to nature itself as a given for guidance. I would like the critics to show what alterations are generally good and what are generally wrong? I would especially like them to show what standard to employ to make this judgment.

Let me give some examples of this problem. In a classic essay against cloning Leon Kass, a conservative critic of redesign and former chairman of the President's Council on Bioethics, argued that cloning was wrong because it was unnatural asexual reproduction.[32] Supposedly it severs the natural connection between sex and reproduction. A close reading of the essay, however, shows that the problem is not reproduction without sex. It is sex without its natural end-reproduction. Reproduction has been the ambit that has guided and structured our most powerful passion for millennia. Cloning is a deep threat to this natural limit. But, so too are all forms of artificial birth control, especially the pill, which only requires one partner to be involved in reproductive decision making. Does Kass really wish to condemn the use of the pill?

With respect to the so-called natural desires described by Arnhart, if we understand them as they are described they cannot state universal moral desires. That they happen frequently I have no doubt. That they should be thought of on the basis of morality *simpliciter* seems plainly wrong, even in terms of the moral sense to which Arnhart appeals. Consider his natural desire to acquire wealth. After stating the obvious that wealth is desired as a secondary good to help our family and friends achieve a good life, Arnhart announces, "Wealth is also desired as a display of status or prestige."

Since Arnhart believes that his "natural" morality is also expressed in the great moral systems worldwide I would ask him to cite any such moral system that holds that the ostentatious accumulation of wealth as a symbol of status is a good or a noble thing.[33] Robin Leach may point to something widespread and titillating but do we really want to think of it and do our moral systems point to it as a display of virtue?

The same point may be made more boldly with respect to war. Though I am a strict pacifist on theological and political grounds, I do agree that war is a persistent feature of human nature. On just war terms, war is thought moral to defend the rights of the innocent, including ones near and dear. For Arnhart human beings desire war when fear, interest, or honor moves them to fight. I dissent on two grounds, one theological and one general. As a theologian I submit that sin is the genesis of war. People who do not really believe that He rose on the third day are prone to fight for others or themselves because they believe that this may be the last time they will ever see them or be alive themselves. The general objection may be more telling for some. The Western intellectual tradition has a rich discussion about the moral ambit for war. Both the justice of going to war and the justice of fighting war have been deeply and widely thought through.[34] But one searches in vain for a major thinker for whom honor, personal or political, was a just ground for war. Does Arnhart really wish to hold that the tendency to fight over slights to "honor among men" is the basis of virtuous activity, as against virtually the whole of Western moral tradition?

Improvement or enhancement can only be understood as enhancement toward some goal regarded as good or noble. To recognize an alteration one only needs to recognize that which is normal or a norm in around which most members of a species will cluster. Enhancement, however, implies going beyond the norm in a positive direction. What is this positive *telos* to which enhancement tends? It cannot be really derived from nature, for nature is what is being enhanced toward some end that goes beyond what we think of as natural. It is impossible to read the goal out of the process because we want to judge various procedures in light of their capacity to bring us closer to the goal. Hence, I believe, naturalism is ultimately unsustainable in the age of biotechnology. Thus, transnaturalism, that is, theology, must supply the missing piece of the puzzle to provide a comprehensive view of biotechnology.

The question that critics of theology always demand at this point asks, how would biotechnology look different through the eyes of faith? Some critics like Arnhart in this volume believe that the Bible has little if anything of value to tell us about biotechnology and that what it does say is either so general and so much mirrored by philosophic tradition, like a concern for justice or the welfare for others, that the Bible has nothing unique or important to say about biotechnology. I disagree. If the Bible is read after the fashion of a law book or as a fundamentalist looking for proof-texts the paucity of specifics about most crucial issues is easily apparent. The Bible, however, is more than just a collection of fundamentalist proof-texts and theology; while grounded in scripture it is more than just a list of propositions held by the community like the 39 Articles or the Canons of the Synod of Dort.

As I argued at the outset, theology is a statement of the cosmic place of human existence. It is a vision of our nature as divinely begotten, our moral duties of love empowered by God, and especially our eternal destiny. Viewed as such, theology has a crucial role to play in current discussions of biologically based technologies. I wish to offer three examples.

Longevity Research

The first example is longevity research, which has certainly caught the fancy of the public. Certainly, this sort of research is based on two entirely natural and reciprocal passions: fear of death and desire to go on living, especially living in a state of reasonable health.

We might at the outset think of two stages of longevity research. The first involves technologies that are available now to slow aging, decline, and disease. The second involves biotechnologies being developed or planned for dramatic extension of the life span out as, some visionaries claim, to eternity. Some visionaries like Ray Kurzweil and Terry Grossman see the process as involving a blending of the two stages. The subtitle of their new book *Fantastic Voyage* is *How to Live Long Enough to Live Forever*. The general idea is to use to the fullest extent currently available nutritional strategies, hormone therapy, early disease detection, and environmental toxicology, so that one can slow the aging and disease process to such an extent that we should be in good health and good spirits when the more radical life extending and life enhancing technologies become available over the next couple of decades. Kurzweil and Grossman dispute the validity of the old proverb about the certainty of death and taxes. They recognize as long as government exists there will be taxes. But they disagree about the necessity of the death part of the proverb.[35] It may very well be that these are rhetorical ploys useful in selling books for a general reader. Their claims may be hype without substance. However, given our natural passion for living, shouldn't we try and wouldn't we want to make this vision a reality?

One might say that this scenario is not possible or highly unlikely. But that only tells us that it would be a waste of resources to try to radically extend our lives. But given our native fear of death this is a hard case to make. One might claim that this does not actually change human nature. This too is problematic. If anything were at the core of human nature it would appear to be a fear of death, which has been the central engine of politics and technology for a millennium. Its corollary, the desire for survival, is exactly what Darwinism teaches is the engine of evolution.[36]

If it is a bizarre idea or a misuse of resources, as I submit it is, it is not because nature teaches us so. It is because faith makes fear of death per se seem completely unreasonable. I do not need to spend resources on the pursuit of near immortality that could otherwise be used to feed the hungry or treat the addicted because I am already going to live forever in God's provenance and in the circle of His love. As a serious Christian I am not commanded to live forever. Such a life is a given and one does not command that which will happen anyway. I am commanded to steward the earth's resources to feed the hungry, lift the downtrodden, and transform the lives of prisoners. On this view money spent on the senseless quest for immortality is a profoundly unjust use of resources. Hence, theology provides precisely a reflection on the technological pursuit of immortality that at least agrees with many individuals' feeling that this technology would be at least bizarre. But theology goes deeper, giving us a cosmic context for concluding that such a pursuit is at the very deepest level faithless and immoral.[37]

Psychopharmacology

The second example is the growing capacity of pharmacology to alter our sense of pleasure about our lives and about the persons and events that populate our lives. I shall use this as one example of a much larger set of issues raised by the possibility of altering either our intentional or natural responses to events and thoughts in general.

Aristotle, the first great biologist, has a rich discussion of emotion or proper feelings as a key to moral judgment. At least part of what moral judgment consists in, for Aristotle, is to have the right feeling of pleasure or disgust in the right amount toward the right person in the right situation.[38] In modern philosophy the key tradition developing the concept of feeling as the center of moral life is that tradition which comes primarily from Hume, through Adam Smith, to Darwin himself and beyond. For Hume when we feel pleasure at the perception of an act or a person we are feeling that the act is good. In reverse the feeling of disgust or displeasure is the feeling of wrongness. For Hume it is not that pleasure accompanies the recognition of moral rightness. The feeling of pleasure simply is the feeling of rightness.[39] The same structure of thinking is found in Adam Smith and with modifications in Darwin. Implicitly in certain passages in Hume and more explicitly with Smith these feelings of pleasure must be those had by an "ideal spectator," that is, a person who is perfectly knowledgeable about the situation, dispassionate, and disinterested, in other words a person who would have the qualities that Christians believe God to have among the other qualities he has.[40]

But what happens to this sort of moral naturalism when we are able to alter a person's sense of pleasure about life with drugs? Old-style trycyclic antidepressants were quite effective for severe depression but they had side effects such as drowsiness and dry mouth that made them medically unadvisable and typically unwanted by those not suffering from major depression. Many clinicians thought that they were unadvisable for those who had mild depression, or dysthyemia. New forms of mood-altering drugs (SSRIs) have few of these side effects. They can and are being used by people with otherwise normal range "feeling structure" to have better feelings about themselves and the world around them. They will feel less disgust than others at some "acts or characters" and more pleasure at others.

These drugs can treat serious depression with less side effects, which is all to the good. They can also increase our happiness in general, a pursuit rooted in our natural desire for happiness. At the very least they can be used to aim at serious alteration of a fundamental feature of human nature, our feelings of pleasure and disgust. Doubters think that this scenario is highly unlikely or almost impossible. Champions and critics often write as if such a move is right around the corner. Fearsome critics sometimes are too credulous about the hype of promoters. But doubters are wise to recall that in most cases of technology including biotechnology what was once thought of as impossible is now commonplace.

The doubters' problem is that on their own analysis the desire that gives rise to these uses of new drugs and still newer ones is perfectly natural. How then can the pursuit of constant happiness be regarded as foolish (which is what they must hold) unless it is per se impossible? This latter impossibility they cannot show. We do not naturally desire unhappiness we desire happiness.

On the doubters' own ground, the desire for pleasure is natural and must be rooted in our biology, especially our neurology. They are Darwinians, not dualists. Dualism would, in a way, ground the claim that redesign with drugs is impossible. But doubters are generally not dualists, so this move is not open to them. Furthermore, we know that these drugs increase the feeling of pleasure had by otherwise normal people so a dualist would solve the problem by denying that any interaction takes place. The "real me," that is, the soul, is not affected by what happens to the body. This would be a very odd dualism indeed.[41]

What the critics of drug redesign must now show is a way of distinguishing between proper uses of these drugs to treat serious depression (a good that they must admit) and improper attempts to alter the "feeling structure" of a person. One can't say it is wrong because it violates a person's given feeling structure, as some given feeling structures are identical to severe depression and leaving them alone can't be good. Moreover, nature aims at pleasure, so one cannot on this account ask nature to show us the limit of its own desire.[42]

A moral standard independent of nature is the only clear solution to the dilemma at this point. A person who is so depressed that she or he cannot reach out in love to others needs treatment often with these new drugs because such reaching out is the moral core of God's intention for humanity. By the same token one who is so broken that he or she cannot feel God's love is also in need of treatment to be able to feel the love, which is God's greatest gift to us. In the same way if a person cannot feel sad at those moral wrongs that are contrary to God's desire for and love of us, such as war, murder, genocide, and the like then that person needs to be able to feel sad at such events even as God himself feels sad at the evil done by his most important creation. Feeling happy or unmoved at all events one experiences cannot be right, not because of nature but because God does not feel happy at everything we do to one another.

Germline Engineering

My third example is the use of germline genetic engineering to alter fundamental features of our existence such as the human life span, mental acuity, body build, and so forth. Over the last fifteen years researchers have tried without success to treat cases of serious human illness such as cystic fibrosis or OT (a liver disease) with what is called somatic cell gene therapy. In this process researchers determine the specific gene that is not working properly and causing the disease. For example, the gene may not be turned on properly and thus not produce the required protein for proper organ function. After identifying the gene involved, a so-called clean or functioning gene is attached to a delivery system, which is typically a virus. The virus supposedly will spread through the organ and "deliver" the functioning gene to the cells where it will integrate with cellular DNA and "cover up" the mistakes of the non-functioning gene. A number of studies have been done since 1990 using this technique on diseases where a specific gene is known to cause the problem. Unfortunately none of the attempts have been successful. For a variety of technical reasons the virus vector has not worked effectively and the new gene has not turned on properly. Major companies who have invested heavily in this research have pulled back from their initial positions.

What is now on the horizon, with proposals on the table at the National Institutes of Health, is a much more radical technique known as germline gene therapy. In this process a clean gene is inserted into a fertilized ovum, using gene-targeting techniques. Insertion at the beginning means that the gene will turn on as a natural result of the differentiation process of the development of the embryo. As liver stem cells are differentiated, for example, the functional gene will be carried along and bring with it normal functional development.

Germline gene therapy has been effectively studied in animals for years and as of this writing there is no reason to think it will not work in human beings. Moral, not technical, objections have delayed the development of this process and were the reason that somatic cell therapy was tried first. Germline therapy has two useful features that give thoughtful people pause. First, germline gene therapy means changing a germline and thus altering a couple's progeny, without the consent of future generations. Secondly, germline gene therapy could theoretically be used to alter complex holistic features like body build or mental acuity.[43]

Some experts caution against using germline gene therapy at this time because we do not know if it will work well in human beings. At this time, we cannot predict with certainty that we will get the result we want and whether we will own the results.[44] Though these objections must be taken seriously, at most they counsel delay at moving forward. They do not amount to a serious per se objection to germline therapy. Nor could they. If the specific gene that results in cystic fibrosis is fully identified, would one want to leave one's progeny with the possibility that they will risk cystic fibrosis in their children, have no children, or restrict whom they can marry. Surely if we can avoid this Hobson's choice for our progeny we would properly want to.

It is also true that most of the conditions that have attracted media attention such as body build, mental acuity, or gender identity, are either polygenic, as in the case of body build, or both polygenic and the result of complex interactions with one's environment, such as gender identity and mental ability.[45] Thus, it is likely to be much harder than a cursory reading of promoters would suggest to alter these human characteristics. The characteristics we seem to care most about will be difficult but not per se unalterable. This is a sound observation but not a full response. The sound observation leads to the conclusion that we should react warily to the claims of promoters of vast redesign of human nature via biotechnology. Unless they can at least show a plausible scenario in which the basis of mental acuity or body build can be radically altered, we should not be swept up in their dreams. In short, don't believe the hype. However, even if the hype is wrong that does not show that, as of now, the scenarios of promoters are impossible. Nor does it show that pursuing these scenarios is wrong except insofar as it is likely to be a waste of capital and expertise for now.

The problem with this position is that it is a holding action not a comprehensive response. The logic of this position is mirrored in the kind of claim that could have been made 150 years ago: "Since we have no idea of how a trip to Mars might be accomplished it would be a waste of resources to attempt it." In the process of technological development and exploration we now know quite precisely how such a trip might be made and what we have to decide is whether it is worth it, not whether it is possible.

To say that germline engineering of what I have called global characteristics is wrong one must employ a standard by which to determine its moral status that is independent of the engineering itself or of calling it a waste of time. The standard ultimately must be based on a teleology of human existence that would be at least a sensible answer to the question: What is the purpose of human existence? Only on such a basis as this can we show that some kinds of global germline engineering are wrong, or even that all are wrong. Severe depression is in many ways a global mental feature that affects memory, mental acuity, human behavior, body mass (particularly obesity) and often can contribute to sleep disorders. Depression of this sort has a serious genetic component as does manic depression and bipolar disorder.[46] I think it is hard to make a case that germline genetic engineering aimed at avoiding this sort of depression from a person's descendants would be wrong. If so, then we require a cosmic or theological context of human existence to show why this is a proper use of germline engineering, and "improving" one's body shape or height cannot be considered as such.

If my view of theology is correct then theology must bear witness to a transcendent truth about the human place in the cosmos, not just give naturalistic reasons for its preferences. What then is the place of theology in the public dialogue about biotechnology? There has been a vigorous discussion in recent years over the place of religion in the making of public policy in pluralistic regimes such as ours. Three general positions have been advanced in the literature.

First is the strict separationist view. On this view religious convictions have no place in either the public desire for certain policies or the actual making of the policies by political representatives. When asked about religious leadership in the antislavery or civil rights movements, representatives of the position reply: "you can get the right result for the wrong reason." These theorists hold that policies should flow from what Rawls in his later work calls an "overlapping consensus" in the public for which policy is made. According to Rawls and his followers this consensus of policies should be based on neutral reasons that exclude religion in pluralistic regimes such as ours.[47]

This view seems to take the "public" out of public policy. If anything is true about the American public it is that they are deeply religious. They do not agree about religion, but they are profoundly religious. What these authors focus on is our deep religious pluralism and the need to find common ground for public policy. The common ground cannot be religion, for that is where our deepest disagreements lie. Rather the ground should be religiously neutral. But this is just as problematic as the position it rejects. Ignoring religion is just as controversial as employing it. Public atheism is no more a neutral ground than its opposite and it is not convincing to religious people who do not share its premise.

A second view admits that ordinary citizens often use religious reasons as one among the many grounds for their policy judgments. Religious convictions about Divine parenthood certainly motivated support for civil rights, against war and abortion. On this view there is nothing particularly odious about citizens using religious grounds to reach conclusions about these topics. Citizens are who they are and to ask a devout Catholic to avoid thinking about his church's opposition to capital punishment when he votes is to ask him to be someone other than he is when he enters the voting booth. To ask this appears to be little more than enlightenment bigotry against religion. This is a position soundly rejected by this second view.[48]

Those who defend this position, however, believe that public officials should be held to a higher standard. After all they represent the whole public, religious and nonreligious alike. They should not base public policy on either religious faith or antireligious animus. Public officials must offer publicly accessible reasons for the policies they support. Of course they may not convince everyone that their policy choices are correct, but what they must do is offer reasons that could convince an otherwise reasonable person to support their position. To oppose gay marriage they should offer reasons, perhaps drawn from biology, not just quotes from a sacred text.[49]

This position, however, asks public officials to do what it admits is not possible for ordinary citizens: to give up their faith when they enter the public square. To ask a Catholic politician who attends mass regularly and confession monthly to say, "But it does not affect my work as governor on behalf of the poor," is to treat faith like a coat that can be taken off to suit the weather. This drastically misunderstands real faith, which defines the core of one's being. It requires of public officials, who are religious, nothing short of hypocrisy if they are to maintain their public position. If his or her faith matters so little, it should be given up completely. But if it really matters to the person's life why should we not want sincere and honorable people to represent us, especially if they are sincere enough to give us a thoughtful accounting of their faith and its intersection with their political life.

The third view is the most hospitable to real religious conviction. As presented by its best advocate, Michael Perry, it only requires religious believers to take religiously based positions either as voters or leaders that are at a minimum "comprehensible" to those who do not share the faith on which it is based.[50]

Take for example capital punishment. Opposition to capital punishment is a decidedly minority position in America. Overwhelming majorities support it. If a Catholic politician opposed it on solid religious grounds all he would be required to do in this view is to state his opposition in a comprehensible manner such as: "I believe that all life is sacred, even the life of the most despised. When someone can be placed in a situation such as life in prison without parole in a way that protects innocent life we may not intentionally take a life as we would be doing with capital punishment." When asked if the pope's office influenced his position he should forthrightly say: "Yes, what of it." Though clearly rooted in his Catholic faith this statement is at least comprehensible to the majority who do not share it.

In another example, consider my earlier discussion of longevity or even "immortality" research. If I were in a position to vote on such funding I would obviously vote against it. My reason would be religious. Immortality has already been given by God at Calvary. Many do not share this view or are doubtful about it. But I submit that even those who do not believe it or are uncertain about it can comprehend it. To clearly argue against a position such as this implies an understanding of the position rejected. Bewilderment, not opposition, is the proper response to what one does not comprehend. Proponents of capital punishment are not bewildered by the position of the Catholic politician. They understand it well. They just do not agree.

Of these three views, I find the third most congenial. But ultimately, I believe that they are all at least partially mistaken. What they all hold is that shaping public policy is a matter of giving reasons for a particular policy position. For the religious this has generally meant giving reasons under the rubric of natural law, reasons based on "nature" that could persuade any rational person.

This move is mistaken. What religious persons should aim at is moral conversion, which is much the same as religious conversion.[51] No religious person I know became a serious Christian because they concluded that Leibniz's modal ontological argument for the existence of God is perfectly valid. This would be a classic Enlightenment argument for faith. Rather, they had a "mighty change of heart" upon hearing a sincere Christian witness mostly from believers who have never heard of Leibniz.[52] People have a serious change of heart about abortion after seeing photographs of the developing fetus, even just the eight-week-old fetus, not from hearing rigorous arguments from people like this writer about how human life begins at conception. The same is true in the case of Mrs. Schiavo that so exercised the nation. People have been affectively moved by seeing the humanity of the handicapped and they worry in cases like that of Mrs. Schiavo about her humanity and the need to protect it.

Like religious conversion, moral conversion in the public square is a matter of the heart, of coming to see the other before us: black, brown, tiny, broken, and sorrowing as we all are as children of a Divine father who has provided for our salvation and happiness—to go beyond ourselves and our cares and care for the weak, the sorrowing, and the criminal. Much as I respect philosophical arguments against abortion and capital punishment, I believe that if we can get others to see the humanity of the weak and broken before us and our responsibility of love to which Christians bear witness, then and only then will we change policy.

Christians must testify or witness to God's love of us and our need to love others along our eternal journey. When we are so sick that we can neither give love nor receive the love of God and others then we need help and the fruits of biotechnology or pharmacology may aid us. But we also bear witness that the reasons mostly given for "enhancements": vanity, domination, "showing off," or plain greed (enhanced players of a sport get bigger contracts) should be rejected by a faithful humanity. Since our destiny and that of those who come after us is secure in His love, we should not turn our backs on Him by spending the precious time and resources we have in vain attempts at biological "improvement" in a world where millions die of malnutrition.

Arnhart: A Response

In his masterful essay in this volume, Larry Arnhart argues that theology cannot solve the issues that have been raised by what I have called strong biotechnology. Significantly, Arnhart does not discuss "theology and biotechnology." Rather, he speaks of "the Bible and biotechnology," as if the Bible was not just a ground of theology, but the whole of theology, which, except for fundamentalists, it is not.

Arnhart offers three reasons for his rejection of a theological response to biotechnology: 1) there is no agreement on the existence or moral authority of God that necessarily leads to "endless debate"; 2) the Bible is unclear and there is no agreement on its interpretation, and, 3) sometimes the Bible teaches immorality and needs to be corrected in light of natural morality. These are serious charges, but each of them can be met.

First, it seems that Arnhart's view requires us to hold in abeyance either the question of God's existence per se, or His giving moral guidance to human beings. But holding that God does not exist, or that he exists in the deist fashion offering no specific guidance to human beings is just as controversial as is theism and presents no solution to the problem of moral and religious diversity. Asking sincere theists to try to solve the sorts of issues that I have discussed without reference to their theism simply won't do. Such a request is equivalent to asking a devoted Darwinist to solve the problem of pain and depression without reference to the Darwinian view that pain or sadness or fear is an evolutionary signal that something is wrong in our lives.

We should not ignore first principles in an argument. It may be plausible, if recourse to a pragmatic solution "for now" can be had, while a broader debate ensues. But when the science is moving so fast and the issues are of such depth and significance as those in longevity research, fundamental mood alterations and genetic transformations that will effect distant generations, the idea that we should just simply avoid discussing first principles, seems a bit like sticking one's head in the sand and hoping one's problems will just go away.

Arnhart's second objection is that the Bible is unclear in many places and its interpretation is highly contested. All of this is true. Over forty years ago a literalist Old Testament scholar concluded on the basis of Exodus 21:22–23 that a relatively moderate position on abortion would be biblically acceptable.[53] In this well-known passage the Torah provides that if during a fight a woman is caused to miscarry then the man shall pay a fine. If, however, the woman is killed then the life of the guilty shall be taken. The implication is that the life of the unborn is not equal to that of the adult. This example is specifically used by Arnhart to show that on key matters that will affect one's judgment on questions of biotechnology such as stem cell research the Bible is a vague and imprecise guide.

There is no doubt that the Bible is a contested text that does not answer twenty-first-century questions with the precision of a literalist oracle. The relevant question, however, is whether nature offers any more precise or uncontested guidance. The answer is that it does not. Many traditionalist conservatives—Catholic, Protestant, Jewish—and philosophers believe that same-sex intimacy is condemned by nature because inherently unreproductive erotic intimacy is inconsistent with the natural, immanent teleology of sexual intimacy that is reproductive.[54] Others read the teaching of nature differently. They point to the pervasive character of homosexuality in the animal kingdom. They also point to sophisticated theories of why homosexuality would persist in nature.[55] These theories require group selection models to make them work. On this issue the debate among those who seek to base at least part of morality on human nature is at least as fierce as between moral theologians or between theists and atheists. Merely appealing to nature does not solve the question. It just transfers it to a new arena.

Consider further the current debate about embryonic stem cell research and the moral status of the early embryo. Many religious and nonreligious authors agree with Arnhart and the authors he cites that stem cell research can possibly do much good in the future and that given the ambiguous moral status of the early embryo, stem cell research is certainly acceptable. Lutheran theologian Ted Peters, for example, has argued that the early embryo has many life-giving potentials: a single new birth, twins, or life-giving stem cell research.[56] Arnhart and Peters may be right, but many other thinkers argue that science (read nature) teaches that human life begins at conception or at least that moral standing does.[57] If the general principle "do not kill" is a teaching of nature, as Arnhart believes it is, and when the two teachings about the embryo and about killing are combined in one fashion or another, we would conclude that nature condemns stem cell research. I will not resolve this debate here I only point out that it appears that nature is no more an uncontested guide to a solution for contested issues than is theology.

The third claim is that the Bible sometimes teaches immoral or morally ambiguous actions that need to be corrected by an appeal to natural morality. It is only in the supposedly true light of nature that we can resolve the morally ambiguous elements of what appears in the Bible. There are at least two responses that the serious theologian can make to this argument. The first is a general response and the second would be a reanalysis of the examples of "biblical immorality" cited by Arnhart.

The general response deserves to go first because in its light we can see the proper theological response to Arnhart's examples. To state the conclusion boldly: Christian theology flows from or is grounded in scripture but not equally. As I argued in the beginning, Christian theology is a teaching of the covenant community that bears witness to its conviction that in the life, passion, and resurrection of Jesus the deepest meaning and eternal destiny of humanity is to be found. Only those sections of the Old Testament / Hebrew Bible that illuminate the teaching of the New Testament should be regarded by thoughtful Christians as having the normative force of the Gospels. All of the Old Testament must be read in light of the New Testament. The Genesis narrative is foundational for Jews, Christians, and Muslims because it teaches the universal covenant with humankind and the origin of Judaism, Christianity, and Islam in Abraham. The prophetic teaching of the Old Testament mutually reinforces and is reinforced by the New Testament. But only the most diehard inerrantist should want to claim that all of the stories contained in the "histories" of the Old Testament are to be regarded as normative by theology. Christian theology has no difficulty in rejecting the examples of anger, vengeance, violence, and slavery seemingly given sanction somewhere within the pages of the Bible. As an aside, I might note that the antislavery movement was largely a movement of Protestant thinkers not one led by or given power by political leaders or natural lawyers, with the exception of Lincoln. None of the great founders who supposedly learned their lessons from nature saw fit to abandon slavery until their deathbeds, and they would have been most uncomfortable with the masthead of William Lloyd Garrison's *Liberator*, the deeply religious Baptist who started the antislavery movement.

Arnhart's most powerful example is the *Akedah*, the famous story of the binding of Isaac. Of all the stories of the Hebrew Bible this one has been seen as most troubling and most revelatory for Jews and Christians alike. Arnhart's argument is that this is a prime example of biblical immorality, which we recognize as such in the proper light of nature. It is this prerecognition of the immoral nature of the sacrifice of Isaac that is the source of the anguished ways in which theologians have tried to make sense of the story. This is not the place to conduct a full and detailed exegesis of Genesis 21 in the context of the Abrahamic narrative as a whole. Especially it is instructive to read the *Akedah* in light of Genesis 18 where Abraham has tested God's concern for human justice and has found it sound. Abraham pleads with God for the safety of those persons in Sodom and Gomorrah who are righteous and God, in a remarkable dialogue, tells Abraham to get the righteous out before the cities are destroyed.

Arnhart's exegesis of the *Akedah* follows the most well-known treatment in Søren Kierkegaard's *Fear and Trembling*, in which Kierkegaard describes Abraham as an example of the true "knight of faith" who forsakes his ordinary moral qualms (the famous "teleological suspension of the ethical") in obedience to divine command.[58] Arnhart believes that Kierkegaard got it right as far as exegesis of the story goes and that this shows the immorality of the biblical narrative. This analysis requires a response.

It seems as though Arnhart, like many readers, stopped his exegesis of Kierkegaard at *Fear and Trembling*. But the awe-full choice that must be made by Abraham is only comprehensible in Kierkegaard's terms in light of the revelatory epistemology presented first in the *Philosophical Fragment* and expanded in the *Concluding Unscientific Postscript* to the *Philosophical Fragment*.[59] On this account human beings are trapped in what is essentially a metaphysical and epistemic conundrum. They can know nothing of certainty because everywhere they turn the ground of knowledge is fluctuating and unstable. It is a quicksand in which they will be swallowed up unless they are saved from it. Even the Socratic solution for Kierkegaard requires another to point out to Meno what is recollected truth. The other who does the pointing, however, is ultimately just another human being like the rest of us. The foundation of his pointing is no more secure than ours would be. The source of our most precious knowledge must come from outside of us since we are weak, vacillating creatures who see only "through a glass darkly." Out of pure love, for Kierkegaard, an incarnational teacher who is both human and divine brings the certainty of the divine to the humanity of the human. Eternity enters time with the advent of the teacher who is at once God and man. For the believer the key is the moment of existential recognition by the student of the teacher as the Teacher. This teacher gives the foundation of divine certainty to the human learner. After the "moment," which is a sort of conversion like that of Paul's, the whole world looks different and one's recognition of one's relation to others, to God, and to the world is fundamentally changed. This is the world of the "knight of faith" who recognizes his own weaknesses and who savors the infinite love of God who would come as teacher to those in darkness. Once one has recognized the teacher and His transcendent goodness and infinite love, the secondary character of our normal forms of value, that is, the "ethical" life of Judge Wilhelm, becomes clear.[60]

If one recognizes such love in the moment then how can it be wrong to make an absolute commitment to the source of this love as Abraham does on Mount Moriah? God has never once let Abraham down. He has called him to found a great people. He has provided for him in times of need. He has listened to and discoursed with Abraham and his moral pleadings for the people of Sodom and Gomorrah. He has even blessed Abraham in the seemingly impossible provision of an heir at an age far beyond what nature allows. Why should Abraham let God down now? In fact, after the moment, not doing what Abraham does would be a faithless denial of what Abraham has experienced of God's goodness. Contrary to Arnhart, I believe that seen in the comprehensive epistemological light of the incarnational love that Kierkegaard teaches, the binding of Isaac shows the power of a truly self-sacrificing love that is the highest form of human morality. "For those who lose their life for my sake shall find it again."[61]

The claim that this is simply an obedience to an arbitrary amoral will is a claim made by those who both do not put the story in the context of the whole of the Abrahamic narrative and who have not found the moment in which they can experience the love of the incarnational teacher. By the same token, those who reject the "knight of faith" with the easy claim that since we can find no instances of an actual knight of faith (a point Kierkegaard himself admits) we may dismiss the teaching as unreal, may as well say that since there are no actual instances of the purely "Socratic" life we may dismiss the teachings of Socrates. If as I believe, the Socratic teaching ultimately is to be rejected in favor of the knight of faith it will be because of some crucial flaw in the Socratic teaching about knowledge, not because in both cases the real does not fully reflect the ideal.

Conclusion

Arnhart's discussion in this volume seems to show his fundamental desire to employ a method that would lead rational people to agree on moral and policy matters regarding biotechnology. I have argued here that I do not believe that his appeal to nature is any better at solving the problems he alleges are found in a theological, or what he calls a biblical, response.

Arnhart's maneuver has been a grave temptation for theology in the modern period, especially when religion speaks in the public sphere. Theologians want to be influential so they avoid controversial theological conclusions and speak with the voice of Enlightenment rationality that can supposedly appeal to reasonable people as such. For example, we get religious opponents of abortion arguing that "science has proven that human life begins at conception" or that homosexual intimacy is wrong because it violates the reproductive mandate of evolution. This is heard even from some who otherwise condemn Darwin to the eternal flames. This appeal to Enlightenment reasoning is mistaken on two grounds. First, it requires theology to ignore its most important wisdom about the eternal destiny, divine framework, and revealed moral pattern for human existence. By adopting this stance, theology gives up its most powerful claim in order to speak like everyone else. But if the core of Christian faith, Jesus, is not "just like you and me" why should a theologian's witness be the same sort as everyone else's? It should not. Theology should bear witness to the world of the wisdom that it alone contains and explicates how this wisdom answers the most pressing problems and deepest longings of humanity. Like Martin Luther King at the Lincoln Monument theology does best when it testifies of its transcendent vision to a weak and doubting humanity.

The example of King points to the second problem with the appeal to what are now called "neutral reasons," which are especially promoted in the later work of John Rawls. Moral conversion as was required in the civil rights struggle or as is desired by opponents of abortion may not ultimately be about giving cogent reasons. Rather it is about changing hearts or altering one's vision of the world and the other who stands in front of us. On his own account, at the end of his seminal talk at the Lincoln Monument, King threw away his nicely prepared speech, which was an Enlightenment recitation of the wrongs done to African Americans and the legal and political remedies required. After throwing away the prepared text he "spoke the words I was given." This is the only part of the speech we remember and it is the part that actually made a substantive political difference. It moved a nation to see the world differently.

Theology has a vision of human existence to which it must bear witness and which will transform individual lives. This vision provides the context in which judgments can be made about the technological transformation of our existence. These are the very judgments that promoters ignore and critics cannot make with their own tools of reason and judgment. It appears that an older tradition remains true: theology is the queen of the sciences.

Notes

1. The list of works that might be referred to here is enormous but two may be cited with confidence. Jaroslav Pelikan, *The Christian Tradition: A History of the Devel-*

opment of Doctrine, 5 vols. (Chicago: University of Chicago Press, 1975–1991); Alister McGrath, *Christian Theology: An Introduction*, 3rd. ed. (Oxford: Blackwell Publishing, 2001).

2. This position is generally known as post-liberal theology. It rejects the liberal view that theology must speak to the world with Enlightenment reason in order to make itself heard. See especially, George Lindbeck, *The Nature of Doctrine* (Philadelphia: Westminster, 1984) and Bruce Marshall, *Trinity and Truth* (New York: Cambridge University Press, 2000).

3. See Nicholas Wolterstorff, *Reason Within the Limits of Religion Alone* (Grand Rapids, Mich.: William Eerdmans, 1984).

4. See David Wiggins, *Sameness and Substance* (Oxford: Blackwell Publishing, 1980). See also Baruch Brody, "A Formal Theory of Essentialism," in Baruch Brody, *Abortion and the Sanctity of Human Life: A Philosophical View* (Cambridge: MIT Press, 1975), 134–149.

5. See Ursula Goodenough, *The Sacred Depths of Nature* (New York: Oxford University Press, 1998); see also Ralph Waldo Emerson, "The Divinity School Address," in *Three Prophets of Religious Liberalism*, ed. Conrad Wright (Boston: Beacon Press, 1961); and see also John Dewey, *A Common Faith* (New Haven: Yale University Press, 1960).

6. Goodenough, *Sacred Depths*, x, 139.

7. Ibid., 101–102.

8. See especially Donald Davidson, "A Coherence Theory of Truth and Knowledge," in *Truth and Interpretation*, ed. E. Lepore (Oxford: Blackwell, 1986), 307–319; and see also Donald Davidson, *Inquiries into Truth and Interpretation* (Oxford: Blackwell Publishing, 1984). Consider also Willard Van Orman Quine, *Word and Object* (Cambridge, Mass.: MIT Press, 1960).

9. At this point I follow the great narrative theologians Alasdair MacIntyre and especially Stanley Hauerwas. See Hauerwas's *Character and the Christian Life*, 2nd ed. (New York: Macmillan, 1990). See also Stanley Hauerwas and David Burrell, "From System to Story," in Stanley Hauerwas et al., *Truthfulness and Tragedy* (Notre Dame, Ind.: University of Notre Dame Press, 1977), 15–39.

10. Luke 1:50.

11. Matthew 1:20; Matthew 3:17; Luke 9:35;Mark 9:7; John 1:1–5, 1:14; Matthew 9:6; Luke 5:24; Luke 7:34; John 3:13; Luke 11:30; John 10:29–30; John 6:69; John 3:16. There are over seventy places in the New Testament where the "son of man" formulation in used. The "son of man" terminology comes from the apocalyptic text Daniel. See Daniel 7:13.

12. Matthew 26:39–42.

13. Thus I am not taking sides on the question of whether an afterlife must include bodily resurrection or whether immortality of the soul will do. On the question of personal identity, which is crucial here a useful review is found in Amélie Oksenberg Rorty, ed., *The Identities of Persons* (Berkeley: University of California Press, 1976). The two sides of the resurrection debate are represented in Peter Geach, *God*

and the Soul (London: Routledge, 1969), speaking for resurrection, and Stewart Sutherland, "Immortality and Resurrection," *Religious Studies* 3(1968): 77–89 for immortality of the soul.

14. Matthew 27:37–39; Matthew 25:31–40; Luke 10:30–37.

15. See Edward Vacek, *Love Human and Divine: The Heart of Christian Ethics* (Washington, D.C.: Georgetown University Press, 1996). See also Steven Post, *Unlimited Love: Altruism, Compassion and Service* (Philadelphia: Templeton Foundation, 2003) and Søren Kierkegaard, *Works of Love*, trans. Howard Hong and Edna Hong (Princeton: Princeton University Press, 1998).

16. Aristotle, *Nicomachean Ethics*, trans. J. O. Urmson and J. L. Ackrill (Oxford: Oxford University Press, 1980), IV.2.

17. See Larry Arnhart, *Darwinian Natural Right: The Biological Ethics of Human Nature* (Albany: State University of New York Press, 1998); and William Casebeer, *Natural Ethical Facts* (Cambridge, Mass.: MIT Press, 2003).

18. Consider David Hume, *A Treatise of Human Nature*, ed. L. A. Selby Bigge (London: Oxford University Press, 1888); Adam Smith, *A Theory of the Moral Sentiments* (Indianapolis, Ind.: Liberty Fund, 1982); and Francis Hutcheson, *Essay on the Conduct of the Passions and Affections* (Indianapolis, Ind.: Liberty Fund, 2002).

19. See Lee Silver, *Remaking Eden: How Genetic Engineering and Cloning Will Transform the American Family* (New York: Avon Books, 1998).

20. See Gregory Stock, *Redesigning Humans: Our Inevitable Genetic Future* (New York: Houghton Mifflin, 2002).

21. See, for example, Dan Brock et al., *From Chance to Choice: Genetics and Justice* (New York: Cambridge University Press, 2001); Erik Parens, ed., *Enhancing Human Traits: Ethical and Social Implications* (Washington, D.C.: Georgetown University Press, 2001); Lori Andrews, *Future Perfect: Confronting Decisions about Genetics* (New York: Columbia University Press, 2001); Glenn McGee, *The Perfect Baby: A Pragmatic Approach to Genetics* (Lanham, Md.: Rowman and Littlefield, Publishers, 2001); Leon R. Kass, *Life, Liberty, and the Defense of Dignity: The Challenge for Bioethics* (San Francisco: Encounter Books, 2002); Ron Bailey, *Liberation Biology: The Scientific and Moral Case for the Biotech Revolution* (Buffalo: Prometheus Books, 2005); Ramez Nam, *More Than Human: Embracing the Promise of Biological Enhancement* (New York: Broadway, 2005); and Joel Garreau, *Radical Evolution: The Promise and Peril of Enhancing Our Minds, Our Bodies—And What It Means to Be Human* (New York: Doubleday, 2005).

22. See Stanley Shostak, *Becoming Immortal: Combining Cloning and Stem Cell Research* (Albany: State University of New York Press, 2002). See also Immortality Institute, *The Scientific Conquest of Death* (Buenos Aires: Librosen Red, 2004). For criticism see S. Jay Olshansky and Bruce Carnes, *The Quest for Immortality: Science at the Frontiers of Aging* (New York: W.W. Norton, 2001).

23. Aubrey de Grey's work can be found at http.//www.gen.cam.ac.uk/sens/AdG-Bio.htm.

24. The Transhumanist Association can be found at http.//transhumanism.org. See also Nicholas Bostrom, ed., *How Can Human Nature Be Ethically Improved*

(Oxford: Oxford University Press, 2005) and "Human Genetic Improvements: A Transhumanist Perspective," *Journal of Value Inquiry* 37 (2003): 493–506.

25. See http.//transhumanism.org/index.php/wta/declaration.

26. See Langdon Winner, "Resistance Is Futile," in *Is Human Nature Obsolete?* ed. Harold Baillie and Timothy Casey (Cambridge: MIT Press, 2005), 406.

27. The most well-known version of this view is found in Catholic sources, such as the official Vatican document, *Humane Vitae* (1968). For support, see Germain Grisez, *Contraception and the Natural Law* (Milwaukee: The Bruce Publishing Co., 1964). For a nonreligious argument that looks very much like a Catholic statement see Lionel Tiger, *The Decline of Males* (New York: St. Martins, 1999). Finally, I note the following from the eminent Canadian thinker, not Catholic, George Grant:

> Because sexuality is such a great power and because it is a means to love, societies in the past have hedged it around with diverse and often strange systems of restraint. . . . It is the reversal in the hierarchy of love and sex which has led in the modern world to attempt to remove the relation between sexuality and the birth of children. The love of beauty of the world in sexual life was believed to have some relation to the love of the beauty of the world found in progeny.

See George Grant, "Faith and the Multiversity," in *The George Grant Reader*, ed. William Christian and Shelia Grant (Toronto: University of Toronto Press, 1998), 473.

28. Arnhart, *Darwinian Natural Right*, 12.

29. Ibid., 29–36.

30. See the essays in Erik Parens, *Enhancing Human Traits*. Also consider the literature on man–machine hybrids by writers such as Donna Haraway, Chris Gray, and James Hughes. See, for example, Chris Gray, *Cyborg Citizen: Politics in the Post-Human Age* (New York: Routledge, 2002); and James Hughes, *Citizen Cyborg: Why Democratic Societies Must Respond to the Redesigned Human of the Future* (Boulder, Colo.: Westview, 2004).

31. See Parens, *Enhancing Human Traits*.

32. See Leon R. Kass, "The Wisdom of Repugnance," *The New Republic* (June 2, 1997): 17–26.

33. Arnhart has appealed to C. S. Lewis's *The Abolition of Man* for a statement of a universal ethic based on nature. In an appendix Lewis gathers together from sources worldwide a number of moral precepts. Two things relevant to Arnhart's argument can be noted. First, almost all of the materials that Lewis quotes and of which Arnhart is enamored come from religious texts such as the Laws of Manu, the Bible, or the Analects of Confucius. Secondly, in none of these texts is the accumulation of wealth as a status symbol recommended.

34. On the justice of going to war and the justice of fighting war consider Paul Ramsey, *The Just War* (New York: Scribner's, 1968); James T. Johnson, *Just War Tradition and the Restraint of War: A Moral and Historical Inquiry* (Princeton: Princeton University Press, 1981); and George Weigel, *Tranquillitas Ordinis: The Present Failure and*

Future Promise of American Catholic Thought on War and Peace (New York: Oxford, 1987).

35. Ray Kurzweil and Terry Grossman, *Fantastic Voyage: Live Long Enough to Live Forever* (New York: Rodale Books, 2004).

36. Arnhart, *Darwinian Natural Right*, 31. In the social contract tradition, especially in Hobbes, it is fear of violent death that is the primary engine that drives human beings to form societies as a common defense against those who would create war and murder.

37. This point is made most strongly by Kant, namely, that moral imperatives require freedom on the part the agent to whom the imperative or command is addressed.

38. Aristotle, *Nicomachean Ethics*, II: 6.

39. Hume, *Treatise of Human Nature*, III, I, 2.

40. See Smith, *A Theory of the Moral Sentiments*, 112–114. Consider also Roderick Firth, "Ethical Absolutism and the Ideal Observer," *Philosophy and Phenomenological Research* 12 (1952): 17–45.

41. This sort of dualism simply won't do from a religious perspective. If temporal events have no effect on the "soul" then the soul would have no memory of them. Hence, it seems that any continuity of a "person" from temporal life to afterlife would be impossible.

42. Arnhart, *Darwinian Natural Right*, 6.

43. Consider David Resnik, Holly Steinkraus, Pamela Langer, *Human Germline Genetic Therapy: Scientific, Moral and Political Issues* (New York: R. G. Landes, 1999); and Leroy Walters and Julie Gage Palmer, *The Ethics of Somatic Cell Gene Therapy* (New York: Oxford, 1996). See also the fundamentally important essay by Tris Engelhardt, "Germline Genetic Engineering and Moral Diversity: Moral Controversies in a Post-Christian World," *Social Philosophy and Policy* 13 (1996): 47–62. Englehart argues that reason cannot determine a specific "human nature" that can guide germline engineering. But in a crucial footnote he writes, "The reader should be given notice: The author is an Orthodox Christian who holds that although reason does not provide canonical moral content, revelation does." I of course have tried to expand here on this point. The place where Englehart and I differ is that I believe that theology should bear witness to its truth to the world and try to change what hearts it can.

44. See American Association for the Advancement of Science, *Human Inheritable Germline Modifications* (Washington, D.C.: AAAS, 1999) and President's Council on Bioethics, *Beyond Therapy: Biotechnology and the Pursuit of Happiness* (New York: HarperCollins ReganBooks, 2003).

45. See Philip Kitcher, *The Lives to Come* (New York: Free Press, 1997); and Francis Fukuyama, *Our Posthuman Future: Consequences of the Biotechnology Revolution* (New York: Picador, 2003).

46. Consider Samuel H. Barondes, *Better Than Prozac: The Future of Psychiatric Drugs* (New York: Oxford University Press, 2003); and Paul R. McHugh and Philip

R. Slavney, *The Perspectives of Psychiatry* (Baltimore, Md.: Johns Hopkins University Press, 1998).

47. See John Rawls, *Political Liberalism* (Cambridge: Harvard University Press, 1988); and Robert Audi, *Religious Commitment and Secular Reason* (New York: Cambridge University Press, 2000).

48. See Kent Greenawalt, *Religious Convictions and Political Choice* (New York: Oxford University Press, 1991).

49. Ibid., 231–243.

50. See Michael Perry, *Love and Power: The Role of Religion and Morality in American Politics* (New York: Oxford University Press, 1991).

51. See Walter E. Conn, *Conversion: Perspectives on Personal and Social Transformation* (New York: Alba House, 1978) and his *Christian Conversion: A Developmental Interpretation of Autonomy and Surrender* (New York: Paulist Press, 1986).

52. This is the idea of being transformed or being born again. See John 1:12–13.

53. See Bruce Waltke, "Old Testament Texts Bearing on Abortion," *Christianity Today* (November 8, 1968): 99–105.

54. See Stanton Jones and Mark Yarhouse, *Homosexuality: The Use of Scientific Research in the Church's Moral Debate* (Chicago: Intervarsity Press, 2000) and Robert A. J. Gagon, *The Bible and Homosexual Practice* (Nashville, Tenn.: Abingdon Press, 2002).

55. John McNeill, *Taking a Chance on God* (Boston: Beacon Press, 1996); Joan Roughgarden, *Evolution's Rainbow: Diversity, Gender and Sexuality in Nature and People* (Berkeley: University of California Press, 2004); and Marlene Zuk, *Sexual Selections: What We Can and Can't Learn About Sex from Animals* (Berkeley: University of California Press, 2002).

56. See Ted Peters, *Playing God? Genetic Determinism and Human Freedom*, 2nd ed. (New York: Routledge, 2002), 179–187.

57. Robert George at Princeton, for example, holds that science proves that human life begins at conception. See for example, personal statements by Robert George and Alphonso Gomez-Lobo, *Human Cloning and Human Dignity*, President's Council on Bioethics (New York: Public Affairs, Perseus Book Group, 2002), 294–306.

58. Søren Kierkegaard, *Fear and Trembling*, edited and trans. Howard Hong and Edna Hong (Princeton: Princeton University Press, 1983).

59. Søren Kierkegaard, *Philosophical Fragments*, ed. and trans. Howard Hong and Edna Hong (Princeton: Princeton University Press, 1985) and *Concluding Unscientific Postscript*, ed. and trans. Howard Hong and Edna Hong (Princeton: Princeton University Press, 1992).

60. Søren Kierkegaard, *Stages on Life's Way*, ed. and trans. Howard Hong and Edna Hong (Princeton: Princeton University Press, 1988).

61. Matthew 16:25.

Suggested Further Readings

Andrews, Lori. *Future Perfect: Confronting Decisions about Genetics.* New York: Columbia University Press, 2001.

Arnhart, Larry. *Darwinian Conservatism.* Charlottesville, Va.: Imprint Academic, 2005.

Arnhart, Larry. *Darwinian Natural Right: The Biological Ethics of Human Nature.* Albany: State University of New York Press, 1998.

Bailey, Ron. *Liberation Biology: The Scientific and Moral Case for the Biotech Revolution.* Buffalo, N.Y.: Prometheus Books, 2005.

Brock, Dan, et al. *From Chance to Choice: Genetics and Justice.* New York: Cambridge University Press, 2001.

Buller, David, ed. *Becoming Immortal: Combining Cloning and Stem Cell Therapy.* Albany: State University of New York Press, 1999.

Fukuyama, Francis. *Our Post-Human Future: Consequences of the Biotechnology Revolution.* New York: Farrar, Straus and Giroux, 2002.

Garreau, Joel. *Radical Evolution: The Promise and Peril of Enhancing Our Minds, Our Bodies—And What It Means to Be Human.* New York: Doubleday, 2005.

Green, Ronald M. *Babies by Design: The Ethics of Genetic Choice.* Hartford, Conn.: Yale University Press, 2007.

Green, Ronald M. *The Human Embryo Research Debates: Bioethics in the Vortex of Controversy.* New York: Oxford University Press, 2001.

Harris, John. *Enhancing Evolution: The Ethical Case for Making Better People.* Princeton, N.J.: Princeton University Press, 2007.

Kass, Leon R. *Life, Liberty and the Defense of Dignity: The Challenge of Bioethics.* San Francisco: Encounter Books, 2002.

Kristol, William, and Eric Cohen, eds. *The Future Is Now: America Confronts the New Genetics.* Lanham, Md.: Rowman and Littlefield Publishers, 2002.

Lawler, Peter Augustine. *Stuck with Virtue: The American Individual and Our Biotechnical Future.* Wilmington, Del.: ISI Books, 2005.

McGee, Glenn. *The Perfect Baby: A Pragmatic Approach to Genetics.* Lanham, Md.: Rowman and Littlefield Publishers, 2001.

McKibben, Bill. *Enough: Staying Human in an Engineered Age.* New York: Times Books, 2003

Nam, Ramez. *More Than Human: Embracing the Promise of Biological Enhancement.* New York: Broadway, 2005.

Parens, Erik, ed. *Enhancing Human Traits: Ethical and Social Implications.* Washington, D.C.: Georgetown University Press, 1998.

Paul, Diane B. *The Politics of Heredity: Essays on Eugenics, Biomedicine, and the Nature-Nurture Debate.* Albany: State University of New York Press, 1998.

President's Council on Bioethics. *Human Dignity and Bioethics: Essays Commissioned by the President's Council on Bioethics.* Washington, D.C., 2008: http://www.bioethics.gov/reports/human_dignity/index.html.

President's Council on Bioethics. *Beyond Therapy: Biotechnology and the Pursuit of Happiness,* foreword by Leon R. Kass. New York: ReganBooks, HarperCollins, 2003.

President's Council on Bioethics. *Human Cloning and Human Dignity: The Report of the President's Council on Bioethics,* foreword by Leon R. Kass. New York: Public Affairs Reports, Perseus Books, 2002.

Sherlock, Richard. *Nature's End: The Theological Meaning of the New Genetics.* Wilmington, Del. ISI Books, 2008.

Sherlock, Richard. *Ethical Issues in Biotechnology*. Lanham, Md.: Rowman and Littlefield Publishers, 2002.

Silver, Lee M. *Challenging Nature: The Clash of Science and Spirituality at the New Frontiers of Life.* New York: Ecco Press, HarperCollins, 2006.

Silver, Lee M. *Remaking Eden: Cloning and Beyond in a Brave New World.* New York: Avon, 1998.

Stock, Gregory. *Redesigning Humans: Our Inevitable Genetic Future.* Boston, Mass.: Houghton Mifflin, 2002.

Tirosh-Samuelson, Hava, and Christian Wiese, eds. *The Legacy of Hans Jonas: Judaism and the Phenomenon of Life.* Leiden and Boston: Brill Academic Publishers, 2008.

Contributors

LARRY ARNHART is Professor of Political Science at Northern Illinois University with specializations in the history of political philosophy, history of biology, and ethics of biotechnology. He is the author of four books and many articles. His most recent books are *Darwinian Natural Right: The Biological Ethics of Human Nature* (SUNY Press, 1998) and *Darwinian Conservatism* (Imprint Academic, 2005). Dr. Arnhart is Associate Editor for *Ethics and Biotechnology and Bioengineering* for *The Encyclopedia of Science, Technology, and Ethics* (Macmillan Reference). He is also a member of the editorial boards of *Politics and the Life Sciences* and *Evolutionary Psychology.*

RONALD BAILEY is the Science Correspondent for *Reason Magazine.* His work has appeared in *The Best Science and Nature Writing 2004*, the *New York Times Book Review*, the *Washington Post*, the *Wall Street Journal*, *Smithsonian*, *National Review*, *Forbes*, and many other publications. He edited *Earth Report 2000: Revisiting the True State of the Planet* (McGraw Hill, 1999) and is the author of *ECOSCAM: The False Prophets of Ecological Apocalypse* (St. Martin's Press, 1993). In 1995, he edited *The True State of the Planet* (Free Press). In addition, he has produced several series and documentaries for PBS television and ABC News. His most recent book is entitled *Liberation Biology: The Moral Case for the Biotech Revolution* (Prometheus, 2005).

RONALD M. GREEN is the Eunice and Julian Cohen Professor for the Study of Ethics and Human Values, Chair in the Department of Religion, and Director of the Ethics Institute at Dartmouth College. In 1996–1997, he served as the founding Director of the Office of Genome Ethics at the National Human Genome Research Institute of the National Institutes of Health. He is the author of *The Human Embryo Research Debates: Bioethics in the Vortex of Controversy* (Oxford, 2001) and *Babies by Design: The Ethics of Genetic Choice* (Yale, 2007).

LEON KASS, MD, is Professor in the Committee on Social Thought at the University of Chicago and Hertog Fellow in Social Thought at the American Enterprise Institute. Dr. Kass was the Chair of the President's Council on Bioethics from 2002–2005.He is the author of several books and many articles. Among his books are *Toward a More Natural Science: Biology and Human Affairs* (Free Press, 1984); *The Hungry Soul: Eating and the Perfection of Our Nature* (Chicago, 1994); *The Ethics of Human Cloning* (with James Q. Wilson) (AEI, 1998); *Wing to Wing, Oar to Oar: Readings on Courting and Marrying* (with Amy A. Kass) (Notre Dame, 1999); *Life, Liberty and the Pursuit of Human Dignity* (AEI, 2002); and *The Beginning of Wisdom: Reading Genesis* (Chicago, 2003).

RICHARD SHERLOCK is Professor of Philosophy at Utah State University. His research interests include the conscience and early modern arguments for religious toleration and ethical and conceptional issues in biotechnology, such as "playing God," "genetic trespassing," and "the precautionary principle." He is the author of *Ethical Issues in Biotechnology* (Rowman and Littlefield, 2002) and *Nature's End: The Theological Meaning of the New Genetics* (ISI Books, 2008).

LEE M. SILVER is a Professor at Princeton University in the Department of Molecular Biology, and the Woodrow Wilson School of Public and International Affairs. He is a member of the Program in Science, Technology, and Environmental Policy, the Center for Health and Well-Being, and the Office of Population Research, at the Woodrow Wilson School. His research interests include human genetics and reproduction as well as the social and political analysis of the influence of religion on the acceptance of biotechnology and the influence of biotechnology on our notions of humanity, life, and the soul. He is the author of *Remaking Eden: How Genetic Engineering and Cloning Will Transform the American Family* (Harper Perennial, 1998), published in fifteen languages, and *Challenging Nature: The Clash of Science and Spirituality at the New Frontiers of Life* (Ecco Press, 2006).

SEAN D. SUTTON is Assistant Professor in the Department of Political Science at the Rochester Institute of Technology. He received his PhD from the University of Dallas. He is the editor of *Perspectives on Politics in Shakespeare* (Lexington, 2006). He is currently working on a book exploring the Enlightenment foundations of rational choice theory.

HAVA TIROSH-SAMUELSON is Professor of History, Director of Jewish Studies, and Irving and Miriam Lowe Chair of Modern Judaism at Arizona State University. Her research focuses on medieval and early-modern Jewish intellectual history, with an emphasis on the interplay between philosophy and mysticism. In addition to numerous articles and chapters, she is the author of *Between*

Worlds: The Life and Work of Rabbi David ben Judah Messer Leon (SUNY Press, 1991), which received the Hebrew University award for the best work in Jewish history for 1991, and *Happiness in Premodern Judaism: Virtue, Knowledge and Well-Being in Premodern Judaism* (Hebrew Union College Press, 2003). She is also the editor of *Judaism and Ecology: Created World and Revealed World* (Harvard University Press, 2002), *Women and Gender in Jewish Philosophy* (Indiana University Press, 2004), and *The Legacy of Hans Jonas: Judaism and the Phenomenon of Life* (Brill Academic Publishers, 2008). She is the recipient of the grant for the Templeton Research Lectures on the Constructive Engagement of Science and Religion (2006–2009) for the project Facing the Challenges of Transhumanism: Religion, Science, and Technology.

Index